# BASS STRAIT
## Australia's Last Frontier

Edited by
**STEPHEN MURRAY-SMITH**

Published for ABC Enterprises for the
AUSTRALIAN BROADCASTING CORPORATION
Box 9994 GPO Sydney NSW 2001
150 William Street Sydney



*First published 1969*
*Reprinted 1974*
*Revised edition 1987*



National Library of Australia
Cataloguing-in-Publication entry
Bass Strait, Australia's last frontier.
  Rev. ed.
  ISBN 0 642 52707 5.
  1. Bass Strait (Tas. and Vic.). I. Australian
  Broadcasting Corporation. II. Title.
910'.0916576

*Set in 10.5/11 pt California by Midland Typesetters*
*Printed by Bridge Printery (Sales) Pty Ltd, Rosebery 2018*

# CONTENTS

# THE AUTHORS

**STEPHEN MURRAY-SMITH** is a Reader in Education at the University of Melbourne and editor of the magazine *Overland*. He has written extensively on the history of Bass Strait, has visited many of the islands, and for over twenty years has spent his summers among them.

**JN JENNINGS** at the time of his death in 1984 was Visiting Fellow in the Department of Biogeography and Geomorphology at the Australian National University.

**RHYS JONES** is well-known for his wide-ranging archaeological studies in the Asian/Pacific region. He is Senior Fellow in the Department of Prehistory of the Research School of Pacific Studies at the Australian National University.

**NJB PLOMLEY** is by profession an anatomist. In recent years, however, he has devoted himself to research work on the Tasmanian Aboriginals. His edition of the journals of George Augustus Robinson, *Friendly Mission*, was said to have doubled our knowledge of the Tasmanians, and an edition of the Flinders Island journals of Robinson is pending. Brian Plomley lives in Launceston.

**ROBIN PRYOR**, former academic and UN consultant on migration, is now the Uniting Church minister at Blackburn, Victoria.

**DOMINIC SERVENTY** had a distinguished career in CSIRO, and pioneered the study of mutton-birds. He is retired and lives in Perth.

**ALISTAIR GILMOUR**, former Executive Officer of the Great Barrier Reef Marine Park Authority, is now Professor of Environmental Studies at Macquarie University and Director of the Centre for Environmental and Urban Studies at that university.

**LAURIE HAMMOND** is Director of the Victorian Institute for Marine Sciences, Melbourne.

**JEANNETTE HOPE** holds the position of Palaeoecologist in Archaeology, Department of Prehistory, Research School of Pacific Studies, Australian National University.

**RICK WILKINSON** lives in Melbourne. He is writer for *The Bulletin* and for *Australian Business* on oil and gas.

# PREFACE

In 1969 the Australian Broadcasting Commission, as it then was, commissioned a series of talks on Bass Strait to be broadcast over its Victorian transmissions. The talks were well received and were published as a small book, *Bass Strait: Australia's Last Frontier*. The book itself proved popular, passing through several impressions over the years. In 1984 it was decided that, rather than republish the book again in its original form, it should be revised and brought up to date, wherever possible by the original authors. This present book is the result.

All entries have been revised, most of them extensively, and figures have been brought up to date. JN Jennings' re-writing of his chapter on the geological history of Bass Strait was accomplished immediately prior to his death in August 1984, which we record with regret. CEB Conybeare, author of the original chapter on petroleum exploration in Bass Strait, died some years prior to Mr Jennings. His work is replaced in its entirety by Rick Wilkinson's contribution, and the other completely new chapter is that by Laurence Hammond, on recent and current research work in Bass Strait, a chapter which could not have been written in 1969 but which could not but be written in 1985. Dr Rhys Jones has chosen to extend his previous contribution rather than revise it. Professor Warren does not consider it necessary to summarise the contributions, as he did in the original edition.

We thank the readers who have continued to demand this little book, especially those in Bass Strait itself, and whose support has greatly encouraged this new edition. References to additional material are given at the conclusion of each chapter, while some may care to note that references to over 1500 items on Bass Strait will be found in *Bass Strait Bibliography*, edited by Stephen Murray-Smith and John Thompson, and published by the Victorian Institute of Marine Sciences in 1981.

S.M-S.

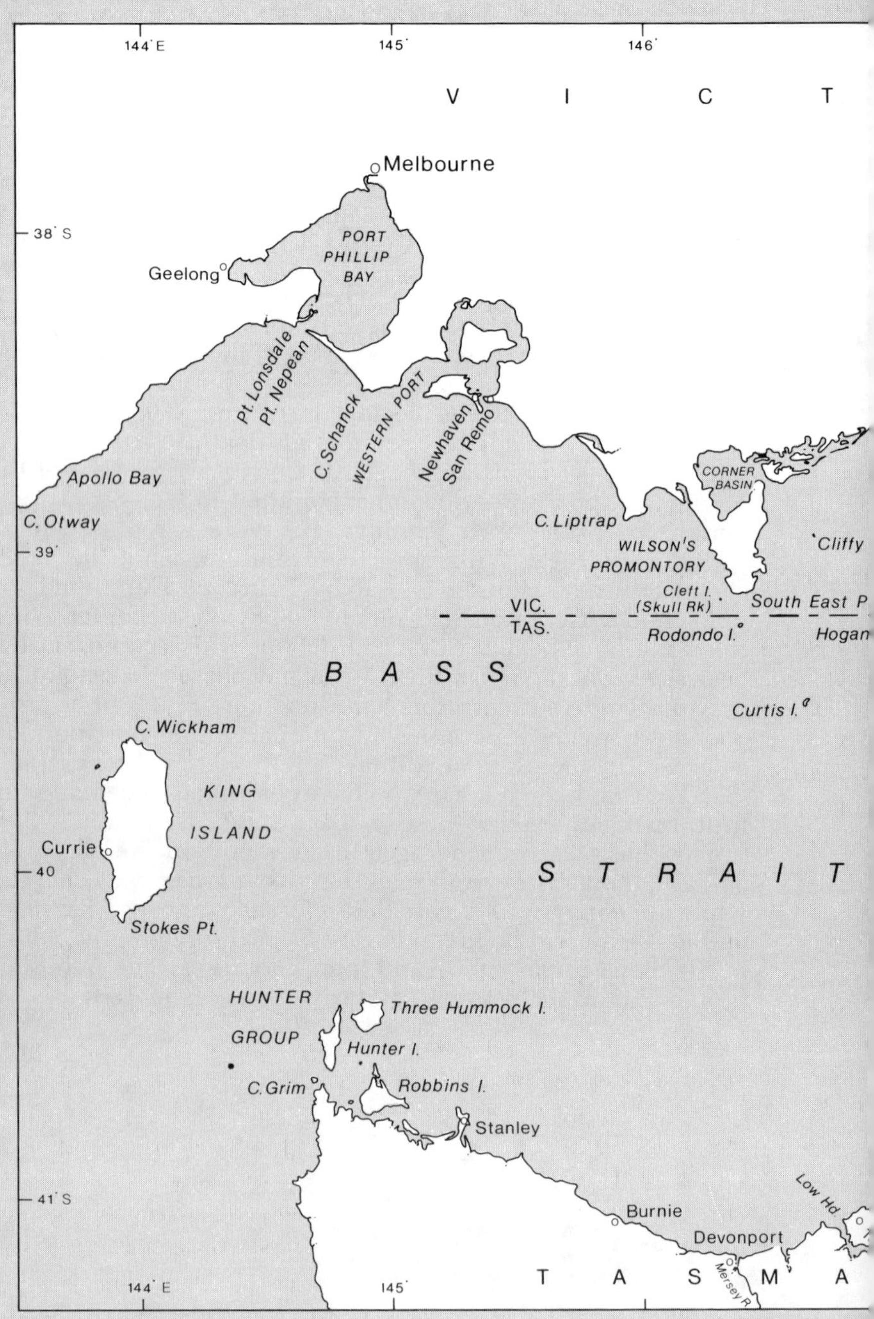
144°E
145°
146°
V I C T
Melbourne
38°S
PORT PHILLIP BAY
Geelong
Pt. Lonsdale
Pt. Nepean
C. Schanck
WESTERN PORT
Newhaven
San Remo
Apollo Bay
C. Otway
C. Liptrap
CORNER BASIN
Cliffy
39°
WILSON'S PROMONTORY
Cleft I. (Skull Rk)
South East P.
VIC.
TAS.
Rodondo I.
Hogan
B A S S
Curtis I.
C. Wickham
KING ISLAND
Currie
40
S T R A I T
Stokes Pt.
HUNTER GROUP
Three Hummock I.
Hunter I.
C. Grim
Robbins I.
Stanley
41°S
Low Hd
Burnie
Devonport
T A S M A
Mersey R
144°E
145°

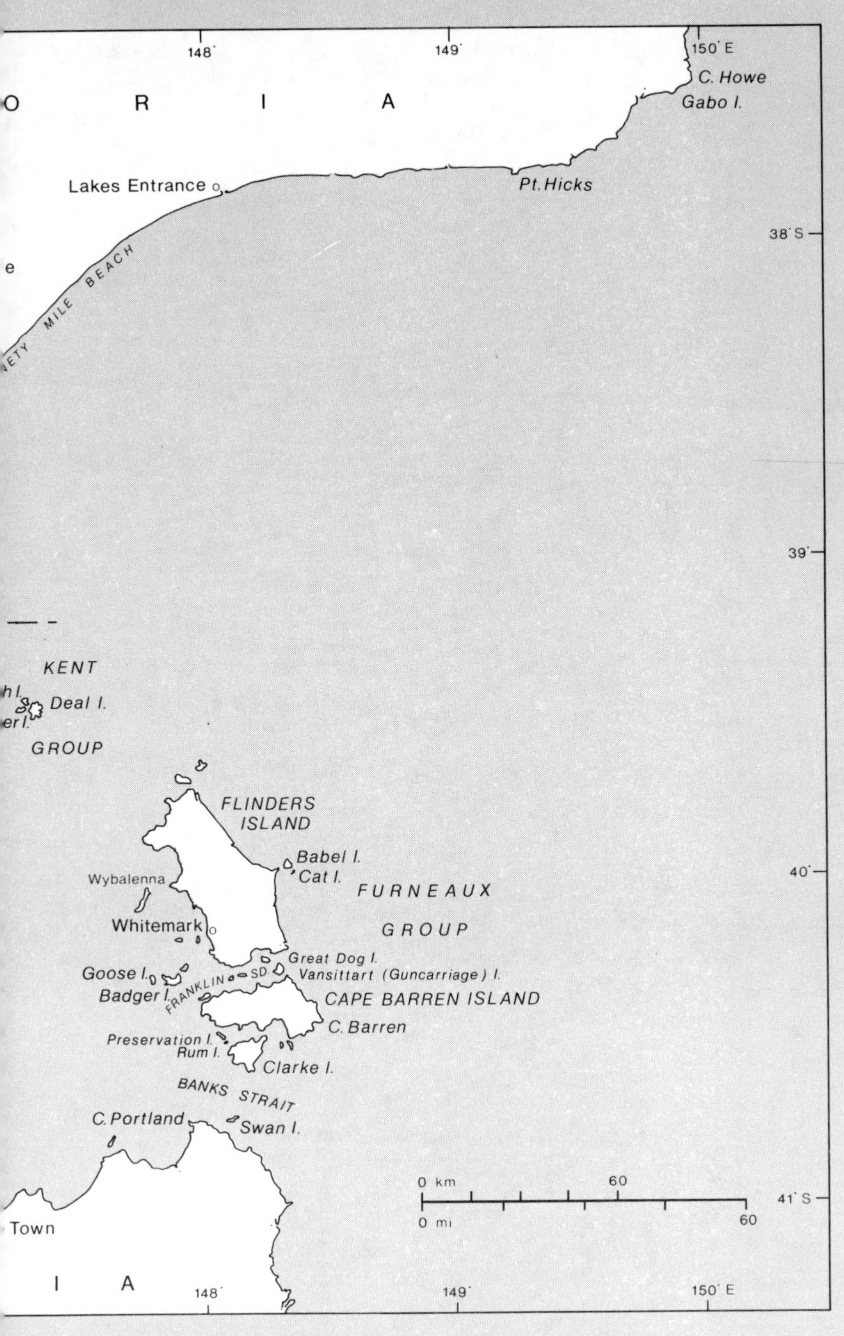

148°
149°
150° E
O R I A
C. Howe
Gabo I.
Lakes Entrance
Pt. Hicks
38° S
NINETY MILE BEACH
39°
KENT
Deal I.
GROUP
FLINDERS
ISLAND
Babel I.
Cat I.
Wybalenna
F U R N E A U X
40°
Whitemark
G R O U P
Great Dog I.
Goose I.
SD
Vansittart (Guncarriage) I.
FRANKLIN
Badger I.
CAPE BARREN ISLAND
C. Barren
Preservation I.
Rum I.
Clarke I.
BANKS STRAIT
C. Portland
Swan I.
0 km
60
0 mi
60
41° S
Town
O R I A
148°
149°
150° E

# 1
# BASS STRAIT: DISCOVERY and EXPLORATION

## Stephen Murray-Smith

Like so many things in early Australia, it started with grog. And there's even an island called Rum Island to prove it. I have flown over Rum Island, a little Pope's nose of an island hard by Preservation Island. I'd like to be able to say that I could still sniff the spirits, rough old Bengal rum it was too, 600 feet up and nearly two hundred years later. But you wouldn't believe me. The best I can do is to say that a local fisherman told me that he'd found a puncheon of rum there. I don't believe *him*.

Rum Island and Preservation Island. They roll off the tongue, those names. They're good names even in a sea of islands where there are many marvellous names, waiting for a poet that they've never had: Tin Kettle Island, Guncarriage Island, Pea Jacket Point, Badger Corner, Killiecrankie Bay, Thunder and Lightning Bay.

Let the rum wait for a minute. Let's go back before the grog. The early explorers, from Tasman on, skirted gingerly around the bottom of what we now call Tasmania. Some asked themselves if it might be an island, but none deliberately sought an answer. The reason was clear enough. After Tasman everyone knew there were good anchorages down south there, and that was the place you made for after a wracking voyage running your easting down over thousands of miles of stormy seas. But once there you wouldn't want to beat back against the westerlies, nor would you want to coast closely up the Tasmanian east coast when a south-easterly, not uncommon, could quickly put you on a lee shore. That's exactly the way that Tobias Furneaux felt about it in 1773. He was captain of the *Adventure*, and Cook of the *Resolution*, on Cook's second voyage. They lost each other somewhere in the south Indian ocean, and didn't meet up again until New Zealand three months later. In the interim Furneaux nearly *did* discover Bass Strait. He got near enough to spy in the distance Flinders Island and what we now know as the Furneaux Group, but he got frightened of the weather and he turned away. The secret was to be a secret for another quarter-century yet.

The First Fleet came round Van Diemen's Land, and so did the Second, and most of the ships that came to Australia, to New Holland, to Botany Bay as they called it, in those first ten years. Those that didn't come south-about came north-about, through the treacherous tropical seas, or perhaps directly across the Pacific from Cape Horn. The key in this 200-mile-wide keyhole was turned by the *Sydney Cove*, Captain Hamilton, master, out of Calcutta with 7000 gallons of rum and other cargo for her namesake destination. She was battered by gales, this unfortunate ship, nearly the whole of her way south and then east. Some of her Lascar crew died manning the pumps, and finally she was blown by an easterly gale through what we now know as Banks Strait, off the north-east of Tasmania, and was run ashore by the 80-year-old captain on a little island. They all got ashore safely, and that is why it is called Preservation.

They might have stayed there a long time, in fact they might even have perished there, but for one of the first and one of the greatest stories of human endurance in Australia. The ship's longboat was sent off, with five white men and thirteen Lascars, to make its way to Sydney. Within a fortnight they were wrecked on the Ninety Mile Beach. There was nothing for it but to walk, to walk over several hundred miles of unknown bush and coast. They were harassed and speared by the blacks, though they were sometimes helped by them too. Within a month, nine of the eighteen had died. Another month, and there were only three left, a European and two Lascars. They were picked up at Coalcliff, south of Botany Bay, and taken by boat to Sydney.

But three was enough. It set the wheels turning. A little vessel called the *Francis* went down to Preservation to rescue the survivors of the *Sydney Cove*. On board was Matthew Flinders, twenty-three years old. While he was away another young man of spirit, George Bass, had also been down in Bass Strait, but this time skirting the northern edges of it. This was Bass's great journey to Westernport Bay in his 28-foot whaleboat, powered by six strong volunteer blue-jackets. It was a voyage, Flinders said many years later, which 'has not perhaps its equal in the annals of maritime history'.

We can see these two young men sitting over their grog when they both got back to Sydney. We even know pretty well what they said to each other. They talked about currents and tides and the run of the swell, and they came to the conclusion that what they had sailed into was not a great embayment but almost certainly a strait, a strait that would cut six or seven hundred miles off the sea route to Sydney, a strait that would be the only bottleneck on the whole run from Europe, in other words a strait that would have incalculable economic and strategic significance. And they talked too about the seals, the fur seals so numerous that the rocks of Bass Strait seemed to crawl, the seals that were to be the first great economic export staple of Australia, in the shape of skins and oil, the seals that were to be to the Australia of the first vital twenty years of last century what the sheep were to become thereafter.

Governor Hunter sent Bass and Flinders south in the *Norfolk* in 1798 to look for their open seaway. Beating down with them went Captain Bishop in the *Nautilus*. Bass and Flinders were looking for fame, but Bishop was looking for money. By the time Bass and Flinders worked their way west to Hunter Island, and saw on December 9 1798, the swells roar on what we now know as Dangerous Bank — 'Mr Bass and myself hailed it with joy and mutual congratulation, as announcing the completion of our long-wished-for discovery of a passage into the Southern Indian Ocean' — Captain Bishop and his men were clubbing the seals and skinning them in their hundreds and in their thousands, and were setting up in Kent Bay, on the south coast of Cape Barren Island, the first settlement in Australia outside the Sydney area itself. Within months adventurers were pouring south to the straits, the Sydney merchants were frantically fitting out ships to reap the fortune awaiting them, convicts with an eye to freedom were thinking that the Bass Strait islands were nearer than China, and those avid moneymakers the Yankee merchants of New England were planning their raids on the Australian sealing islands that were in their time to lead to not a few split heads and diplomatic protests.

So Bass Strait's bridal day had come. But what kind of a bride was she? What *is* Bass Strait? Most Australians these days think of it as a half-hour gutter between the mainland and our island State. There's a lot you cannot see from 30 000 feet.

Bass Strait is not a few islands surrounded by a lot of water. It would be almost as true to say that it consists of a lot of islands in a bit of sea. Not quite as true, really, because when a shark or cray boat does only seven or eight knots it takes seventeen hours to get back home to Port Albert or Port Welshpool from down in the sounds between Flinders Island and Cape Barren Island, and in seventeen hours, or indeed a couple of hours, a placid, easy sea, no more than a little 'jobbley', as I've heard it called, can become a monstrous thing, great waves splintering themselves against the granite rocks and islands, and a screaming

*A small girl places flowers on one of the many graves to be found on the Bass Strait Islands. This is one of a ship's boy buried on Erith Island in 1886.*

12

wind obliterating with drift the sea and the land and the sky itself. No man who's been caught by a sudden change in Bass Strait will ever deny there's a lot of sea there, least of all the fishermen, among whom there are remarkably few fools. They watch the horizon, these men, when they are out there, just as primitive man watched the treeline and the skyline, constantly vigilant.

Well, then, there's a lot of sea but there's a lot of land too, too much land for seamen to trust, as the widows of nearly two centuries could tell us. There are, it is said, 126 pieces of land in Bass Strait. Some are mere rocks poking out of the sea, like Skull Rock which many have seen off Wilson's Promontory, or the prominent Devil's Tower, 363 feet (111m) high in the Curtis Group. Many Victorians too will have seen Rodondo, the big cone of granite six miles off the Prom., which John Bechervaise and his Geelong College boys landed on in 1947, and which few have been on since. There are many less spectacular islands, mere domes of granite rising like migrating whales from beneath the sea. The Hogans, for instance, east of the Promontory; Brian Stackhouse pastures fine cattle here, where you wouldn't think a hare could scratch a living. Round Flinders Island in particular there are dozens of small islands, inhabited once by Robinson Crusoe families, but deserted now, the floors of the old houses crumbling, strange old Bibles sometimes found in dark corners, snakes living in old ovens which once fed a swarm of lively, dark-skinned youngsters. Of all the Bass Strait islands, only six or seven are permanently inhabited.

About three thousand people live in Bass Strait, but many more make their living from it. The 'western straitmen', as they once used to be called, live on King Island, bang in the middle of the western end of Bass Strait, between the north-west horn of Tasmania and Cape Otway in Victoria. Some 2000 Australians live here, raising their cattle and sheep on an island which is physically not dramatic but which has a long history of drama and of horror, which I will talk about later in this series. The other western straitmen live on Three Hummock Island, south of King, two of them farming a large and lovely island: John and Eleanor Alliston who, in thirty years of isolation and hard labor, have made this corner of Bass Strait their own. You can read about it in Eleanor Alliston's books, *Escape to an Island* and *An Island Affair*.

The eastern straitmen are less numerous but more widely dispersed. For here, south and east of Wilson's Promontory, we have the tattered remnants of the old land bridge between the

*Cairns such as this one were built by bluejackets from Royal Naval surveying ships last century. This structure is on the main peak of Dover Island.*

mainland and Tasmania, over which those departed folk the Tasmanian Aboriginals probably migrated. You could almost island-hop on a log across Bass Strait here now, provided you had a good outboard on the back. This is the corridor for light aircraft making their way to and from Tasmania. An airplane comes down over the Promontory, swings out to the Hogans, then over twenty miles of open sea to the lovely Kent Group, with its thousand-foot pink granite cliffs and its sheltered sandy coves. Another leap of 30 miles or so and we are over Flinders Island, and from here we are over land nearly all the way to Tasmania, with the exception of the last nine miles or so over Banks Strait. Never run away with the idea that all these Bass Strait islands are tiny. Flinders is over half a million acres, King is over a quarter of a million, Cape Barren is over 100 000. As many skippers have found, they are big lumps to run into on a dark night.

Flinders Island is not only the biggest but in many ways also the most interesting. Here, in the early days, the sealers and the escaped convicts and the wreckers congregated with their kidnapped native women; here George Augustus Robinson brought the pitiable remnants of the Tasmanian Aboriginals, and here, in the 1830s and the 1840s, at a place called Wybalehna, they nearly all died away. We know they sat on the hillsides looking at the blue shadow of Tasmania on the horizon, weeping; and we can still see here the brick chapel they built with their own hands, and in which they were taught the catechism. In terms of the associations it conjures up, this is one of Australia's most sad and significant historical buildings. When these talks were first published in 1969 it was still being used as a shearing-shed. Today it has been restored as a memorial and it and its surroundings are in the hands, amongst others, of the descendants of the Tasmanians who built it.

Flinders Island, too, is an island of magnificent spectacle. The cloud-topped Strzeleckis, soaring to over 2500 feet (777m), dominate the south of the island, and are greeted over the few island-studded miles of Franklin Sound by the almost equally impressive peaks of Cape Barren Island. This has been the homeland for generations of the islanders whom the anthropologist Norman Tindale called 'a whole new human population brought into being by hybridisation'. There are some thousands of these people in the Australian community today, but only a handful still live on Cape Barren Island. Sprung, in the first half of last century, from sealers in Bass Strait and their Aboriginal women, they evolved into a vigorous and interesting

*Part-Aboriginal school-children, Cape Barren Island, 1969.*

people, at one time great boat-builders and great boatmen. Often dispossessed from their ancestral islands by white settlers, and resented by whites because they refused to accept their values, their revival of their pride in their heritage has been a notable feature of the past twenty years.

Over the beautiful, island-studded Franklin Sound from Cape Barren Island lies the great mass of Flinders Island, home to over a thousand Australians: farmers, fisherfolk, teachers, builders and all the callings that go to make up an independent community. There are few rich people and no 'great houses' there, and comforts and security have had to be struggled for over difficult generations, and isolation is always a problem, but few who live there would wish to leave.

Bass Strait is a frontier where men and women and children live a Hebridean existence between land and sea. Fishing and farming often go together, and so does a resourcefulness and a tenacity which it is good to see. These islands are tough to master, but this means that they haven't yet been spoiled. Perhaps here at least in Australia we will find the wisdom to preserve the character of the land and life even if we are destroying it elsewhere.

**Further reading:**

[Bishop] Francis Nixon: *The Cruise of the Beacon* (London, 1857)

[Canon] Marcus Brownrigg: *The Cruise of the Freak* (Launceston, 1872)

Charles Barrett: *Isle of Mountains* (Melbourne, 1946)

Patsy Adam Smith: *Moonbird People* (Adelaide, 1965)

Eleanor Alliston: *Escape to an Island* (Melbourne, 1966)

Patsy Adam Smith: *There Was a Ship* (Adelaide, 1967; new edition, Melbourne 1983)

JS Cumpston: *First Visitors to Bass Strait* (Canberra, 1973)

Stephen Murray-Smith: *Beyond the Pale: the Islander Community of Bass Strait in the Nineteenth Century* (reprinted from *Papers and Proceedings*, Tasmanian Historical Research Association, vol 20, no 4, December 1973)

Mavis Thorpe Clarke: *The Hundred Islands* (Sydney, 1976)

Stephen Murray-Smith (ed): *Mission to the Islands: the Missionary Voyages in Bass Strait of Canon Marcus Brownrigg 1872-1885* (Hobart, 1979)

RM Fowler: *The Furneaux Group Bass Strait* (Canberra, 1980)

Jim Davie: *Latitude Forty: Reminiscences of Flinders Island* (Melbourne, 1980)

Eleanor Alliston: *Island Affair* (Melbourne, 1984)

Ida West: *Pride against Prejudice: Reminiscences of a Tasmanian Aborigine* (Canberra, 1984)

Stephen Murray-Smith: 'Islands of Bass Strait' in Geoffrey Dutton (ed.), *The Book of Australian Islands* (Melbourne, 1986)

# 2
# The Geological History of Bass Strait

## JN JENNINGS

Tasmanians think that Tasmania is very different from the other island to the north, and in some ways this is true. Geologically it is not so, because the rocks and the way they are arranged tell us that Tasmania was built as part of the mainland and that their long histories have marched in common. So the foremost question to put about Bass Strait is — when did it form? When did it cut off this fragment of the continent?

The answer isn't simple, nor do we know it fully, although oil exploration has greatly improved our knowledge of the Strait's earlier history. In one sense we can say that Bass Strait formed only a few thousand years ago; in another that it is tens of millions of years old. Even this great age makes it a very young feature geologically, because Australia as a whole is an old continent.

For much of its history the geographical features of south-eastern Australia ran north to south; the seas and mountain ranges which formed and disappeared many times stretched from Tasmania through Victoria into New South Wales and sometimes into Queensland. The geographies of those times bore little resemblance to that of today. Indeed in the latter part of this history, for most of the time since about 350 million years ago, south-eastern Australia was land, land extended unbroken across the position of Bass Strait. More than that, the whole formed part of the huge former continent of Gondwanaland, which included South America, Africa and Antarctica as well as Australia. What are now submarine rises and ridges beneath the Tasman Sea were still fused to this major land mass east of Tasmania, Victoria and New South Wales.

Then about 140 million years ago (at the beginning of Cretaceous time) came the first signs of separation of a future

Tasmania as part of the breakup of Gondwanaland. In southern Victoria from Gippsland to the Otways, an east to west trough developed as a result of faulting across the previous north to south trends of the older rocks. This was the eastern end of the main rift, which grew from west to east and heralded the separation of Australia from Antarctica. At this stage it looked as if Tasmania was going to be left behind as part of the southernmost continent and become simply a land for penguins and seals. However, at much the same time, there was also the beginning of a great NW-SE fault west of King Island and Tasmania, which was eventually to separate them from Antarctica. Parallel with this, the central part of Bass Strait began to sink in a basin elongated in extension of the Tamar valley of Tasmania. The northward movement of Australia was stretching the crust and causing this subsidence at right angles to the pull from the north-east.

These troughs were filling up with sediment as they subsided but the nature of the sediments and the fossils in them tell us that they were being laid down by rivers and in lakes. The east-west trough, comprising the Gippsland Basin in the east and the Otway Basin in the west, chiefly received sediment from the Victorian ranges, whereas the Bass Basin of the NW-SE trough did so from all sides but mainly from Tasmania, streams carrying sediments north-westwards. There was no question yet of the sea's breaking through. The faulting which allowed the basin to sink also permitted volcanic magma to penetrate to the surface and pour out as lava and explode in showers of ash, which were incorporated in the basin fill.

The situation began to change towards the end of Cretaceous time, about seventy to eighty million years ago, when the sea began to trespass into the western end of the trough across Victoria where the Otway Ranges now lie. The Tasman Sea began to open up at this time as elongated masses of continental crust drifted eastwards. These movements folded and faulted the sediments that had accumulated in the trough in Gippsland, though the sea did not invade here yet nor into the Bass Basin.

Early in the succeeding Tertiary Period, a great, deep trench had been created between the mainland and Antarctica, and now the Tasman reached the shelf off Gippsland. By about forty-five million years ago (at the end of Eocene time) the sea had reached shallowly into King Bay, and even penetrated as a narrow estuary into the Bass Basin from the northwest. There was ocean between Tasmania and Antarctica, as yet comparatively shallow.

During the following period of the Oligocene there came a big invasion of the sea, merging the Bass Basin first of all with the Southern Indian Ocean to the west. But by about twenty-nine million years before present, not only was there a deep ocean trench south of Tasmania but the Bass Strait seaway had come into being, with the link to the Tasman Sea also established. Thus over one hundred million years had gone to its formation; such a long period of gestation is usual for major geographical features.

In these last stages of this first creation of Bass Strait marine sedimentation was widespread in the area. Earth movements were dying down and erosion less; coarse sands and silts gave way to finer sediments and biogenic deposits. Indeed, in the following Miocene Period the seas reached farthest into Victoria and onto the present Bass Strait islands, depositing marls and limestones from clearer waters. When they lapped back from this farthermost extension, the geography of Bass Strait was well roughed out.

Nevertheless the geological story of Bass Strait is far from finished at this point, because other factors than those of crustal movement and sedimentation came to play the most powerful role. In the latter part of the Tertiary Period, and even more in the last 2.3 million years of the Quaternary Period, world climate changed more markedly and rapidly than it had done before. We know now that there were glaciers in Antarctica before Bass Strait was opened for the first time. But it was the comings and goings of the vast icesheets of the Northern Hemisphere that is crucial for our present concern with Bass Strait.

Each time these ice sheets grew large, huge amounts of water were locked, frozen, on the continents for thousands of years. This water came from the only large reservoir there is on earth, the oceans; it was evaporated from their surfaces, blown inland, condensed to snow and piled up to form ice. As a result the level of the oceans fell. We do not know precisely how far they fell, but certainly sometimes to the order of 100 to 140 metres. From the evidence of the icesheets of North American and Europe, it used to be thought there were four such ice ages, with consequent lowerings of the ocean level. Now, with much greater knowledge from deposits on the sea floor and in lakes, it is clear there were many more substantial climatic cycles than that, and more ice ages in the Quaternary. So Bass Strait was made land again and Tasmania rejoined to the mainland a number of times in the last two million years.

In these ice ages or glacial stages the whole of Bass Strait became land again. Animals could migrate across it and plants could spread. I shan't concern myself with these biological consequences, only the environmental ones. For example, at these times the rivers which drained into Bass Strait deepened their valleys near the old coasts; thus, borings at Melbourne have shown that the River Yarra cut channels well below sea level, as much as fifty metres down. These rivers flowed over the floor of Bass Strait. There are river gravels west of Flinders Island deposited by small rivers from that island which extended over the sea bottom. The former channel of the River Mersey can be traced to a point about nineteen kilometres off the Tasmanian coast and to sixty metres below sea level.

The deepest parts of Bass Strait, over ninety metres deep, lie between Cape Otway and King Island. When sea level was very low, the Yarra must have flowed south-west to this. There is also a central depression in the Strait, stretching northwest from northern Tasmania, the present manifestation of the Bass sedimentary basin. The River Tamar must have run along this depression to join the Yarra or rather — as Professor Sam Carey of Hobart has cheerily pointed out — for the Yarra to join it. So Tasmanians can fairly claim that the Yarra was a tributary of the Tamar, just as Melbourne started out as a suburb of Launceston!

As yet it hasn't proved possible to trace the former courses of these big rivers across the bottom of the Strait, which over large areas is flat and featureless to an exceptional degree. These rivers must have fashioned channels in it, but Bass Strait is a very rough sea; wave action and currents have eroded here, and deposited sediment there, to smooth it all out since.

When the ice sheets retreated and melted away in warmer climatic stages of the Quaternary, meltwater returned to the oceans and sea level rose in varying degree. How much depended on the length and warmth of these phases. The ones which became at least as warm as now are called interglacials, whereas less complete recoveries are called interstadials. In this complex sequence of climatic events, sea level moved up and down and was liable to stand still at all sorts of levels in consequence. For instance, virtually all the way round from Cape Otway to Wilson's Promontory there is a submarine break of slope between fifty-five and sixty-five metres down. This represents an old coastline where waves cut into the rocks to produce a sea cliff with shore platforms in front. It must belong to a colder time than now, possibly to a lesser glacial maximum.

In the interglacials, sea level rose not only to its present level but sometimes went above it. In the coastlands of Victoria and Tasmania, and on some of the Bass Strait islands, evidence of this was left behind in the form of marine shell beds, sea caves, sea cliffs, and coastal dunelines well above present sea level. Thus beneath Mowbray Swamp near Smithton in north-western Tasmania, there are marine sands about fifteen metres above sea level, which are more than 34 000 years old and probably belong to the last interglacial. There were even higher sea levels; in the big opencut scheelite mine on King Island, beach boulder beds over thirty metres up were uncovered. If in some interglacials the climate became somewhat warmer than at present, glaciers would have been reduced below their present volumes and the oceans would have risen above their present level. However, the farther we go back in time, the more movements of the land are involved as well, and the more difficult it is to disentangle the one cause of changing sea level from the other. In New South Wales, emerged Quaternary sea levels have been found only up to about five metres above the present sea level and it is probable that these higher Bass Strait levels are the result of the tilting up of the crust there.

As the sea rose from low levels, it flooded the incised valleys of coastal rivers, turning them into deep coastal inlets like those of the Tamar and of the Mersey at Devonport. Or else these deeply cut valleys were filled up with estuarine and deltaic deposits in the course of sea level rise, as in the case of the River Yarra.

Each time sea level fell again, fresh erosion took place, removing much of these sediments, so that the features and deposits we can study best are those produced since the last ice age in the last great rise of sea level which brought it to its present resting place. We know most about this particular rise, not only because its results survive more completely, but because it took place within the range of time in which the radioactivity of organic carbon, in such things as fossil plants, shells, bone and even soil humus, enables us to date events more precisely than in any other part of geological time. Such materials found on the sea bottom, or in excavations for docks and bridges from many parts of the world, have given us a good idea of what happened.

The last time the ice sheets and glaciers grew really big culminated about 18 000 years ago, and sea level fell to somewhere around 120 metres down. From that time to about six or seven thousand years ago, it rose rapidly, practically to its present level. Belonging to this long rise, there is the stool of a river red gum which was found nineteen metres below low water mark in a buried channel of the Yarra at Melbourne. The radiocarbon age of the tree was nearly 9,000 years; the coast lay lower than that at the time. Maybe related to this is an old soil and ground surface beneath Port Phillip Bay to a depth of eighteen metres. Sea shells in sediments on top of this soil have been dated 6000 years old, so the soil is older than that; but by how much, we don't know.

Though there is controversy as to what happened in detail, sea level has always been very close to its present stand in the last six or seven thousand years. So throughout this time Bass Strait has been virtually just as it is today. The important question is — when did the land link with Tasmania end? When did the sea break through to form Bass Strait for the last time? This event must have fallen within the time when men have lived on this continent.

Before trying to answer it, I need to tell you a little more about the topography of Bass Strait. I have said already that the deepest part lies north of King Island. But between that island and north-western Tasmania there are a number of reefs, islets and islands, which stand on a broad ridge in the sea floor. Also, on the eastern side of the central depression, there is a whole chain of islands, large and small, between Wilson's Promontory and north-eastern Tasmania, including the Furneaux Group. Again these are the projecting parts of a broad underwater ridge. Granite rocks loom large in the make-up of this ridge, from Blue Tier in Tasmania to Wilson's Promontory itself; many of the smaller islands are entirely of granite. These two ridges and the central depression between them all follow the trends of fractures or faults and are essentially due to the structure of the crust.

As sea level rose in the last 18 000 years, it advanced north-east between the Otways and King Island quite early on. Next it spread south-eastwards along the central depression. Then came the most important happening of all; the sea lapped over the eastern submarine ridge between Flinders Island and Wilson's promontory, cutting Tasmania off from the mainland. Later on, King Island was separated from north-western Tasmania, and even later the sea penetrated between the Furneaux Islands and north-eastern Tasmania to form Banks Strait.

As yet we have no direct evidence when these land links were severed. We can only make inferences with a reasonable degree of probability. Let us take the most important case, between Wilson's Promontory and Flinders Island. Here, the threshold, or sill, which had to be drowned, is a broad area between fifty-five and sixty-five metres in depth between the Kent Islands and the Promontory. It is very flat, but rising abruptly from it are various islands such as the Hogan Group. Some of the islands have what are called 'plunging cliffs', that is, the cliffs do not end close to sea level as is usually the case because wave action has trimmed them back at that level, but continue steeply below the sea down to a former level of wave erosion now submerged.

So this level of fifty-five to sixty-five metres is critical to this question of the final severance of Tasmania from the mainland. Before we can infer when this occurred, we must consider what may have happened to this sill since it was drowned. I haven't previously mentioned tidal action, but this is not a factor to be neglected. Tidal currents can scour deep holes in the sea floor or prevent them being filled up with sediment. Such scour holes are in fact found in the narrow straits between many of the Bass Strait islands. There is a particularly deep one, over 150 metres deep, between West Sister Island and the northern end of Flinders Island. However, the broad sill that matters has not been lowered in level by such scour holes, so that for our present purpose this current action may be neglected.

Currents can also deposit sediment and shallow the sea. Again there is evidence of such action in Bass Strait. For example, Beagle Spit is a long shoal of streamlined form running west to east from the northern end of Flinders Island, and there is a similar feature east of Clarke Island on the northern side of Banks Strait. These shoals have shallowed the sea nine and eighteen metres respectively. Nevertheless there are no such shoals between the Kent Group and Wilson's Promontory.

Then there is the question of whether the land itself has moved up or down or has tilted in these last 18 000 years. Undoubtedly there has been some crustal movement in this time. South Gippsland has had more earthquakes than any other part of Victoria, and earthquakes are just the incidents in human lifetime of the protracted, cumulative crustal movements of geological history. Nevertheless, over the period of time we are considering, earth movements seem to have been quite small in Bass Strait. So the level of the sill cannot have changed much since it was drowned.

Finally, there is also some uncertainty about the time at which the sea, in its last great rise to its present position, was at these depths of 55 to 65 metres. Dated indicators of former sea level from such depths are few. Thom and Roy (1983) have assembled what there are from southeastern Australia, and from their graph of depth against age the closest limits that can be put on the time when the sea reached this level in its last rise, and overtopped the sill of Bass Strait for the last time, are 12 000 and 13 500 years before the present. This may sound a long time ago to people who think in terms of Ancient, Medieval and Modern History. To the historian of nature, Bass Strait came finally into existence only yesterday, and this of course helps to explain the close similarities in some respects, for instance in animal life, between the island state and the mainland. Behind that late event there lies as we have seen a long sequence of happenings in which such tremendous factors as the breakup of the super-continent of Gondwanaland and the kaleidoscopic climatic history of the Quaternary — the Great Ice Age — have played critical roles.


**Further Reading:**

JL Elliott, 1972: 'Continental drift and basin development in south eastern Australia', *J Aust Petrol Expl Ass* 12, 46-51

JN Jennings, 1959: 'The submarine topography of Bass Strait', *Proc Roy Soc Vict* 71, 49-72

JN Jennings, 1971: 'Sea level changes and landlinks', pp 1-13 in *Aboriginal Man and Environment in Australia*, edited by DJ Mulvaney and J Golsen, ANU Press, Canberra

RB Leslie, HJ Evans and CL Knight, eds 1976: *Economic Geology of Australia and Papua New Guinea, 3, Petroleum, Aust Inst Min Metall*, Memo Ser 7:
Pp 67-82, Brown, BR 'Bass Basin, some aspects of petroleum geology'.
Pp 34-41, Colman JAR: 'Gippsland Basin, onshore'
Pp 82-91, Ellenor, DW 'Otway Basin'
Pp 41-67, Threlfall, WF, Brown BR and Griffith, BR: 'Gippsland Basin, offshore'

BG Thom and PS Roy, 1983: 'Sea level change in New South Wales over the past 15 000 years', pp 64-84 in *Australian Sea Levels in the Last 15,000 Years: A Review* edited by D Hopley, Department of Geography, James Cook University, Mono.Ser., Occ.Pap.3

# 3
# BASS STRAIT IN PREHISTORY

**Rhys Jones**

## I — THE PREHISTORIC SETTING

Bass Strait has influenced the history of men living in its proximity for thousands of years, in two distinct and even contradictory ways. I can best introduce these influences by recounting two quite trivial experiences, which for me, sharpened my appreciation of them and their role in the prehistory of the area.

My first experience was flying over the strait in an aircraft. The journey was an easy one lasting about an hour, yet as I sat there fifteen thousand feet above the sea, sipping my coffee, I was crossing one of the most effective geographic barriers in human history, a barrier which once had separated men for over 10 000 years.

Let us cast our minds back some 15 000 years. Below us, where the sea is now, is a broad plain with some high hills rising from it. To the south lies a mountainous peninsula, its centre locked in an ice sheet. We can see rain-bearing winds coming up from the south-west, banking up swollen grey clouds full of snow against the western mountains; and freezing conditions affect the high ground for over half the year. The hills are treeless, and the forests are mostly confined to a narrow strip along the coasts.

Man arrived on the Australian continent at least some 30 000 years ago, having somehow managed to cross the water barriers from Asia. In late glacial times he was well established across the inland plains, and on the south-east coast of New South Wales, too, he was occupying the middle reaches of small coastal valleys a few miles from the sea. South in Tasmania, we have, as yet, no evidence of man's late glacial occupation, but it is highly likely that here, too, he lived along the foreshores and on the coastal heaths.

As the ice age waned and the ice sheets of the world melted, the water ran into the sea, which, between 20 000 and 6000 years ago, slowly rose 300 feet to its present level. Great areas of coastal marsh and plain became progressively flooded and the flat isthmus connecting Tasmania to the mainland became

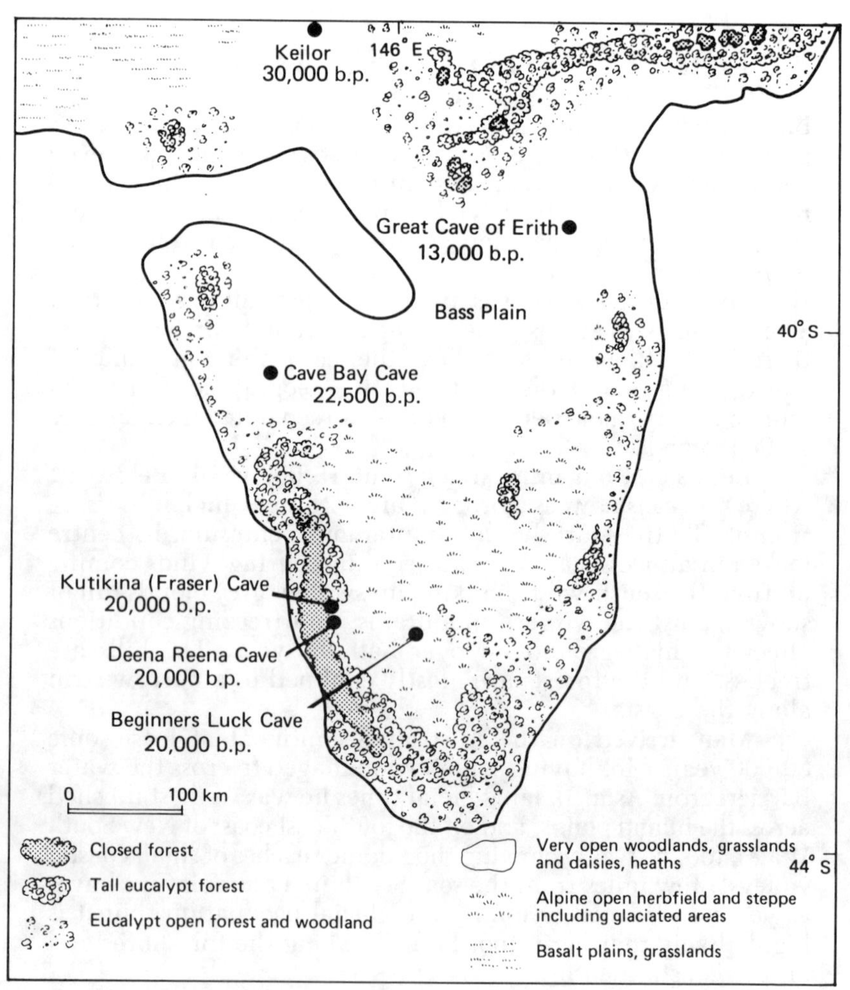


*Environmental reconstruction of the Bassian Plain and 'Tasmanian' area of southeastern Australia at the height of the last Ice Age, about 18 000 years ago.*

a shallow and ever-widening strait about 12 000 years ago.

Tasmania was once more an island, and some people became separated from their fellows on the mainland. From then on, the Tasmanians remained cut off from the rest of the world for 10 000 years, probably one of the longest periods of isolation ever recorded in human history.

The stone tools of the Tasmanians are similar to those recovered from some old archaeological deposits on the mainland. They consist of concave and nosed scrapers, high domed pieces called 'horse-hoof cores' and retouched pebble choppers. In Tasmania, this stone tradition was maintained throughout the entire archaeological sequence for 8000 years until European contact, although there are minor variations from place to place on the island. On the mainland, however, this industrial tradition persisted only some five to seven thousand years ago, when new fashions of stone-tool making swept the continent. These new traditions, which were probably introduced from Asia, specialised in the manufacture of small composite tools, which included points, adze flakes and tiny geometric microliths. None of these influences ever reached Tasmania.

When the Tasmanian Aborigines were first described by the Europeans, the ethnographers noted the simplicity of their material culture. They had no spear-throwers, boomerangs, shields, complex baskets or hafted edgeground axes. It is likely that most of these artifacts had been diffused to, or invented on, the mainland after the formation of Bass Strait, although ancient edge-ground axes have been found in Pleistocene cave deposits in northern Australia.

There were no dingoes in Tasmania at the time of contact, and none have ever been found in fossil deposits. Dingoes are common on the mainland, and their bones have been discovered in caves back to about 7000 years ago. Mainland Aborigines often kept dingoes as pets or as hunting aids, and it is believed that the dogs arrived in Australia with man, as a domestic or semi-domestic animal. It is likely that these events occurred after the severance of Tasmania from the mainland, and that the water barrier of Bass Strait stopped the southward diffusion of the dog. That dogs can be extremely useful to hunting man is attested by the speed with which the Tasmanian Aborigines themselves adopted and trained European dogs once they were available to them. GA Robinson records meeting a group of twenty-five Tasmanian Aborigines in 1831; they had a hundred dogs with them.

Can we then think of the Tasmanian Aborigines as static fossilised representatives of early Australian man and his culture? I think not, for the archaeological sequences on the island show that the Tasmanians discontinued several cultural traits during their long stay on the island.

In a coastal cave which I excavated at Rocky Cape, in north-west Tasmania, the lowest deposits contained numerous remains of scale fish, for example, parrot fish. These deposits, dated from 8000 years ago to about 4000 years ago also contain dozens of fine bone points manufactured from wallaby fibulae. After this period, from three and a half thousand years ago to the ethnographic present, fish dropped out of the diet, and bone tools were no longer manufactured.

In answer to the question 'why should this have happened?' all I can say is that I don't know, except that again I suspect the influence of Bass Strait. The archaeology of remote islands often documents this 'loss of useful arts'. Is it the result of isolation and the lack of outside stimulus? Or is it that cultures as well as animals, stranded on islands, are protected from the full blast of the forces of natural selection which are operating elsewhere?

The isolation of the Tasmanian Aborigines on their island may help to explain certain physical differences, such as in the form of the hair and the face, between them and the mainlanders. Genetic divergence can easily occur between populations isolated from each other for a long time, and this is particularly the case when we are dealing with small numbers of individuals.

Thus the first great influence of Bass Strait on the history of peoples living on its shores has been as a barrier, as a *divider* of cultures.

*

My second experience was holding on to the mast of a small fishing boat trying to edge around Wilson's Promontory in a freshening on-shore breeze. All around me were albatrosses wheeling, and in the water I could see seals and penguins bobbing up and down. My childhood fantasies of the wild southern oceans and the 'Roaring Forties' came back to me. We prudently retired to a small sheltered cove and went on shore. Here I could have sworn that I was in Tasmania. There was a dense scrub of ti-tree, banksia and the grass tree *Xanthorrhoea*; a couple of wallabies broke cover and, between the tides, I could see the same species of shell fish as can be found on the island coast some 200 miles to the south.

The coastal environments of Victoria and Tasmania are

similar to each other, and they share a large number of species of plants and animals. The food refuse left by the Aborigines reflects this similarity. In terms of the foods eaten, the coastal shell middens of Victoria resemble those of Tasmania far more than they do those of the New South Wales coast near Sydney and further north.

In my sites in north-western Tasmania, about half of the protein food was provided by rock platform shell fish such as the turbo *Subninella undulata*, the abalone, the rock welk *Dicathais textilosa* and the limpet *Cellana solida*. Identical shell fish species were eaten on the coast of Victoria, for example, at Wilson's Promontory, in prehistoric times; and, indeed, given shell samples from middens in both areas, I could not tell which came from where. In Tasmania, an extremely important source of protein came from the hunting of seals; on the north coast predominantly fur seals; and on the west coast the now locally extinct Southern Elephant Seal, *Leonina mirounga*. Distribution studies show that prehistoric sealing was carried out along the west coast, and on the north-western and north-eastern corners of Tasmania. On the mainland there is evidence of Aboriginal sealing having been carried out on the Coorong in South Australia, along the Victorian coast and in New South Wales as far north as Sydney. The sealing population which provided this food was based on Bass Strait and, despite the depredations of the European sealers, it is still established there today.

Sea birds such as penguins, large gulls and even albatrosses, provided food on both sides of the Bass Strait. Mutton birds, whose remains are common in shell middens in southern New South Wales and Victoria, were one of the prized foods for the Tasmanian Aborigines, who used to cross treacherous straits to islands such as Hunter, Three Hummocks and the Doughboys in north-west Tasmania, to collect the birds and their eggs.

In the tropical and desert areas of Australia, recent work has shown the great importance of vegetable food in Aboriginal diet, some authors claiming that plants constitute up to eighty per cent of the total food. As we move south into the higher latitudes, the variety and abundance of edible species diminish, and entering the zone of the 'Roaring Forties' the importance of vegetable food in Aboriginal diet was markedly reduced. Most of the plant food was provided by the kernels of the grass trees, and the young shoots of cutting grasses, bracken and other ferns. In Tasmania, the giant kelp was beaten and braised before being eaten.

This diminution of the abundance of vegetable foods in the

high latitudes is a world-wide phenomenon. For both plant and animal foods we can talk about a 'Bass Strait province', characterised by an abundance of marine foods such as molluscs, marine mammals and sea birds; and by a relatively paucity of vegetable foods.

Men living on opposite shores of Bass Strait took advantage of the available natural resources of the area. Thus we find that, though separated for 10 000 years, and in many ways culturally quite different, the coastal Aborigines of Victoria and northern Tasmania were eating the same food, were practising the same economy and as far as middens go, left similar remains for the archaeologist to recover.

In this sense, I see Bass Strait as a unifying force in the history of the Aborigines.

We are left with a paradox: Bass Strait, the geographical divider of men, and Bass Strait the ecological unifier. It is the interplay of these two factors which gives a fascination to the prehistory of the region.

The forces of history seldom stop dead at the whims of colonial enterprise, and it is interesting to speculate whether or not Bass Strait will continue to influence man in the next 15 000 years in the same way as it has in the past 15 000 years.

# II — COLONISTS OF THE BASSIAN PLAIN

The above was written in 1969, only sixteen years ago. Re-reading it underlines what an advance in our knowledge about the prehistory of this continent has occurred since then. There, I recounted two personal experiences which had influenced my archaeological imagination; in this one I shall begin with another.

It was in January 1978 on Erith Island, one of the isolated Kent Group situated in the centre of the eastern gate of the strait. Erith is separated from Deal Island by a kilometre wide, 30 fathom-deep, channel called Murray Passage. To get to the group required a ten hour, stomach-churning voyage from Port Albert, and as we left the lee of Wilsons Promontory we were in the grip of a relentless swell from the south-west. The cliff citadels guarding both sides of Murray Passage was our target, and inside it was this miracle, a sunny calm haven with sandy coves on both shores backed by granite cliffs, while outside was the wild seascape of the strait. Stone tools had been found on Erith Island in 1974. Having been intrigued by this discovery, I was trying, with fellow archaeologist Ron Lampert to get to a huge cave on the northern edge of the island.

The problem was access. We had tried once with a dinghy from the sea, but had almost foundered on rocks. The other approach was by foot along the base of a line of sheer, ten metre-high cliffs. The critical area was a space some thirty metres long, where a slippery downward sloping ledge was swept by each wave even at low tide. After careful counting of the time between each wave, it was possible to dash between them to the safety of the jumble of rocks beyond. Further along, one reached the base of a talus of boulders and earth which spilled almost vertically from the mouth of a gigantic cave. Its floor was twenty metres above the sea and, to reach it, one had to climb tight against the cliff wall where the loose slope deposit met the hard rock. The mouth itself, consisting of great overhanging buttresses of smooth granite, led into a wide chamber almost fifty metres deep, its flat floor consisting of a soft orange deposit. We carried food, water and firewood into this cave and camped there for a few days, being probably the first people to have done so since the end of the last Ice Age.

We excavated a small test pit carefully sieving all the deposit, which within the top 1.20 m, consisted only of sterile water-sorted grit. Then we came suddenly down onto a different unit of brown soft clay and rocky rubble. Within it were large fragments of bones of wombats and wallabies, some heavily burnt so that they were calcined blue grey at the edges. We believe that these animals had been brought as game into the cave and cooked and eaten there by people during a brief visit. Tiny pieces of soft charcoal have been radio-carbon dated, indicating that this event probably took place some 7000 to 9000 years ago. Beneath this unit was another metre, at least, of grey clay and rubble containing thousands of bones of rats and marsupial mice, probably the prey of owls. A date of some 13 000 years ago was obtained from here. The species recovered indicated the nearby presence of several terrestrial ecosystems, including dry grassy plains and diverse shrublands. Pollen analysis from these deposits has confirmed this palaeo-ecological picture. While we were digging towards the base of our excavation a great storm blew in from the south-west, the cave being in the lee of the island, but such was the force of the wind that, blowing over the rocky bluffs above, it doubled up on itself and forced 'willy-waws' or water spouts, and sheets of spray, backwards towards us. While the rocks below boomed under the pounding of the waves, in my mind's eye the sea disappeared and the cliffs of Erith stood as a massive tor rising out of the

*Headland, Erith Island, Kent Group. The Great Cave of Erith, and the loose talus leading to it, may be seen as a dark patch in the cliff-face not far from the point.*

The Great Cave of Erith, in the Kent Group, Bass Strait. Archeological work in this cave is referred to in part two of Rhys Jones's article in this book. The pit dug for archeological research may be seen, left centre. The large mound in the centre of the photograph, with a dish on it, is a natural, largely organic deposit.

cold sandy plains of the last Ice Age. Where Murray Passage now is, then was a mysterious dark cleft splitting the mountain block, its flanks clothed in patches of dense eucalypt and banksia scrub. Owls swooped overhead picking up their prey, and darted into the cave. Far to the west, plumes of smoke indicated the passage of bands of hunters foraging on the edges of low-lying salt marshes on their way to the ice-capped mountains of the south.

Returning to reality, later on the same expedition, we found stone flakes and scrapers eroding out of a wind blown sand dune perched high above what is called the 'Swashway', a tidally inundated pass between Erith and Dover Islands. Charcoal from the artefact-bearing layer has been dated to some 9500 years ago. When Flinders visited and named these islands in 1798, he found them uninhabited and later said of them that 'An extreme degree of sterility seems to prevail through the whole of this small cluster of isles; and they seem to be shunned by almost every kind of animated beings'. Equally all of the large islands in the centre of Bass Strait, such as King Island and those of the Furneaux Group, as well as Kangaroo Island off the coast of South Australia, were uninhabited when first visited by the first English and French expeditions of Flinders and of Baudin. yet by now stone tools and other evidence of prehistoric occupation have been found on every one. How we nowadays explain this paradox would have left old Matthew Flinders, that practical, empirical man of the Age of Enlightenment, quite speechless with disbelief.

With the onset of the coldest phase of the last Ice Age the polar ice sheets increased greatly in size, and water was withdrawn from global atmospheric circulation, resulting in a drop of the sea level. For Bass Strait, the crucial depth lies between minus sixty-five metres and fifty-five metres, since this would mean the difference between a strait sixty kilometres wide and a plain sixty kilometres wide. This 'Bassian Sill' is situated between Wilson's Promontory and the northern coast of Flinders Island, with only Rodondo and the islands of the Kent, Curtis, and Hogan groups always remaining exposed. The sea dropped to this level about 24 000 years ago, thus exposing the floor of the previous strait as a potential road to Tasmania. At 18 000 years ago the sea level was probably 150 m below its present one, exposing the entire width of the continental shelf from west of King Island to the steep underwater canyons east of the Furneaux chain. This Bassian Plain was exposed to a cold, dry and windy climate. Southern Australia in the grip of the last

Ice Age was a cold, dusty place with the great anti-clockwise continental dune system re-activated, sending active dune fields as far south-east as central Victoria and past Canberra into the upper reaches of the Shoalhaven. These dunes, aligned from north-west to south-east, snaked their way along the floor of the Bassian Plain, and they can still be seen fossilised along their original orientations in north-eastern Tasmania as a series of parallel ridges and swales, now vegetated by coastal heaths. In the very sump of the plain, some fifty kilometres north of Burnie, there may have been a great swamp or lake into which drained most of the northern Tasmanian and southern Victorian rivers. It probably balanced this inflow by evaporation in the same way that Lake George in New South Wales still does, since no great river gorge has ever been found cut into the King Island-Cape Otway submarine limestone ridge, the only feasible outflow to the west. The vegetation on the plain probably consisted largely of a dry grassland, similar in some ways to that still existing in a relic form in western Victoria near Mount Arapeles.

As explained in my 1969 essay, it had long been assumed that people first got to Tasmania across this glacially exposed plain, but the first direct proof for this came in 1974 from Sandra Bowdler's excavation in a cave on Hunter Island north-west of Tasmania, and finally in Tasmania itself in 1976 by Albert Goede, in Beginner's Luck cave in the Florentine Valley, a western tributary of the Derwent close to Mount Field West.

The basal dates at Cave Bay Cave on Hunter Island shows first human occupation at about 21 000 to 23 000 years ago, a date not significantly different from that of the first effective exposure of the Bassian ridge as a road to Tasmania. In the Florentine Valley site, the date is 20 000 years ago, and here there are stone tools associated with the bones of wallabies. However, there still remained doubts whether or not the inland regions, as opposed to the coastal ones, of the Tasmanian peninsula were occupied systematically during the full phase of the last Ice Age. These were, however, dispelled by the discoveries from 1981 to 1984 in the Franklin Valley system of south-west Tasmania of numerous limestone caves, some containing evidence of intense prehistoric occupation. At Kutikina (formerly Fraser) Cave on the Franklin, there were superimposed layers containing large numbers of stone tools and burnt and broken animal bones. The basal date from this site was 20 000 years ago, the top one about 14 000 years ago, the occupation spanning the coldest phase of the last glacial maximum. Similar dates have been recovered from half a dozen other sites from

.the same area. At this period the Antarctic ice sheet was only about 1000 km to the south, and there were valley glaciers within a few kilometres from the sites themselves. The people hunted red-necked wallabies, from an environment consisting of an alpine steppe-land very different from the present one of closed temperate rainforest. They manufactured stone tools both from local cobbles of quartzite and also some from material which they obtained from a splatter of naturally produced glass on the western rim of the Darwin Crater, formed when a great meteor smashed into the earth, and in the energy of its impact melted some of the mother rock. This crater, first located scientifically only in 1974, had already been used by the first inhabitants of the Franklin and Andrew Valleys of western Tasmania some 16 000 years ago. They carried the glass to make their tools at least forty kilometres, or three days' walk, from the source. I have elsewhere noted that these inhabitants of south-western Tasmania were at that time, during the height of the last Ice Age, the most southerly human beings on earth.

I would guess that they may have occupied the most southerly valleys of Tasmania only on a seasonal basis, perhaps as summer hunting trips. Their total seasonal range may have extended onto the floor of the Bassian Plain or even to what is now the mainland of Victoria, as winter quarters. In this context it is significant that the final dates of occupation of all of the cave sites so far investigated in the Franklin Valley system fall into the time slot of between 12 500 and 14 000 years ago. This corresponds very closely with the time of the flooding of the Bassian Plain, especially its western part during the global warming at the end of the ice age. The land bridge to the Australian mainland was finally severed at about 12 500 years ago, and this may have totally dislocated the economic system of the hunting societies operating across it. At any rate, there was no further occupation of the inland limestone caves of south western Tasmania, and the old deposits of camp site and hearth became entombed in cold white stalagmite as the rainforest closed in and sealed off this remote country until a few years ago. Quite literally, the Franklin was a lost valley of the ice age, its Pleistocene human history integrally bound up with that of the drowned Bassian Plain, 200 kms to the north.

As the sea rose, the low-lying parts of the Bassian Plain became flooded, but what about the hills cut off once again to form the island chains both to the east and west of it? Archaeological evidence gained during the past decade has shown that human societies occupied these places as the sea rolled back up to its present level. On King Island in 1979 I found, by accident, stone tools eroding out of a sand dune overlooking high cliffs on the southwest coast. Carbon recovered from the occupation level showed that people were living here about 7500 years ago. This was a time after the severance of King Island from Tasmania, although the sea had not yet reached its present level. This was also the case with Erith Island, as described at the start of this paper. It appears that some people lived here briefly at a time when the islands of the Kent Group had been cut off both from Wilson's Promontory to the north and Flinders Island to the south. At this period, in attempting to reconstruct the exact geographies of islands, the situation is complicated by the probable load of sand which the sea was carrying up with it as it rose across the Bassian Plain. In the initial phase of sea level rise, this load may have been too great for the sea to remove for a while, and so it is my guess that sand spits or other features may have existed, temporarily linking landmasses which now are separated by water. Conversely, the sea has also since reaching its present level, scoured some deep erosion channels into the sea bed whch may not have existed at the time of the initial formation of the islands.

Whatever the complexities, it appears that both King Island and Erith Island lost their human inhabitants soon after the islands were formed. However in the case of Flinders Island, as with Kangaroo Island and Hunter Island, people stayed on them until the sea reached its present level and had stabilised there. They lived off shellfish and other coastal resources, leaving well-defined middens. Wayne Orchiston and his team have found fascinating shell middens on the north and west coasts of Flinders Island dating to 6000 years ago, while Ron Lampert has documented a prehistoric occupation of Kangaroo Island until probably some 5000 years ago. Yet finally all of these islands lost their human populations. Why this occurred is a fascinating problem of human history, whether it happened due to the fact that the isolated populations were too small to be viable in the very long term (as measured in terms of scores of generations), or whether conscious decisions were made to leave due to the lack of adequate social needs in such small communities. Once the islands lost their populations, almost all of them remained de-populated. This was because of the great difficulties and dangers of crossing these waters in the simple craft available to ethnographically-recorded Tasmanian and Victorian Aborigines. It is probable that voyages beyond about five kilometres from the coast were extremely dangerous and seldom carried out. However, during the past two or three thousand years, some such voyages were carried out to re-colonise the closer inshore islands off both coasts of Bass Strait. Hunter Island, having been probably abandoned soon after it was cut off from the Tasmanian landmass was, re-occupied probably seasonally by people from about 2500 years ago. On the other side of the Strait Great Glennie Island, some seven kilometres to the west of Wilson's Promontory, was occupied seasonally from about 1500 years ago.

At the very end of the eighteenth century it was open to people like Flinders, Bass and Peron to explore Bass Strait across space; during the past twenty years, it has been left to our generation to continue the exploration of Bass Strait through time.

**Further readings:**

**The Prehistoric Setting:**

PJF Coutts, 1967, Coastal dunes and field archaeology in S.E. Australia'. *Archeology and Physical Anthropology in Oceania*, 1:28-34

Hiatt, Betty, 1967, 1968 'The food quest and economy of the Tasmanian aborigines' *Oceania*, 38:99-133, 190-219

Jones, Rhys, 1967, 'Middens and man in Tasmania' *Aust. Nat. Hist.*, 18:359 – 364

Jones, Rhys, 1968, 'The geographical background to the arrival of man in Australia and Tasmania'. *Archeology and Physical Anthropology in Oceania*, 3:186-215

**Colonists of the Bassian Plain:**

Sandra Bowdler, 1977: 'The coastal colonisation of Australia', in J Allen, J Golson, R Jones (eds), *Sunda and Sahul: prehistoric studies in southeast Asia, Melanesia and Australia*, pp 205-46 Academic Press, London, New York, San Francisco

Sandra Bowdler, 1984: 'Hunter Hill, Hunter Island'. *Terra Australis* 8, Department of Prehistory, Research School of Pacific Studies, Australian National University, Canberra

JN Jennings, 1971: 'Sea level changes and land links', in DJ Mulvaney and J Golson (eds), *Aboriginal Man and Environment in Australia*, pp 1-13, Australian National University Press, Canberra

Rhys Jones, 1969: 'Bass Strait in prehistory', in S Murray-Smith and J Hope (eds) *Bass Strait: Australia's last frontier*, pps 26-31, Australian Broadcasting Commission, Sydney

Rhys Jones, 1977: 'Man as an element of a continental fauna: the case of the sundering of the Bassian bridge' in J Allen, J Golson and R Jones (eds), *Sunda and Sahul: prehistoric studies in southeast Asia, Melanesia and Australia*, pp 317-86, Academic Press, London, New York, San Francisco

Rhys Jones and RJ Lampert, 1978: 'A note on the discovery of stone tools on Erith Island, the Kent Group, Bass Strait', *Australian Archaeology* 8:146-9

Rhys Jones, 1979: 'A note on the discovery of stone tools and a stratified prehistoric site on King Island, Bass Strait', *Australian Archaeology* 9:87-94

Rhys Jones and Jim Allen, 1979: 'A stratified archaeological site on Great Glennie Island, Bass Strait', *Australian Archaeology* 9:2-11

Kevin Kiernan, Rhys Jones and Don Ranson: 'New evidence from Fraser Cave for glacial age man in south west Tasmania', *Nature* 301:28-32

Ronald Lampert, 1981: 'The Great Kartan Mystery', *Terra Australis* 5, Department of Prehistory, Research School of Pacific Studies, Australian National University, Canberra

Orchiston, D Wayne, and RC Glenie, 'Residual Holocene populations in Bassiania: Aboriginal man at Palana, northern Flinders Island', *Australian Archaeology* 8:127-41

# 4
# ROBINSON'S ADVENTURES IN BASS STRAIT

## The story of George Augustus Robinson and the Tasmanian Aboriginals

### NJB Plomley

To say that George Augustus Robinson was a remarkable man is not to praise him unduly. Like all of us, his character was a blend of the attractive and the unattractive, the good and the bad, but there were also in it features of greatness.

Robinson was born of lower middle class parents on 22 March, 1788, probably in London. Nothing is known of his boyhood. As a young man he was employed on construction work at the Chatham naval establishment and then on the building of Martello Towers on the south-eastern coast of England. These activities seem to have extended from about 1806 to 1812, after which he returned to London. There he seems to have occupied himself as a building contractor in a small way until 1823. On 9 September of that year he sailed on the *Triton* as an immigrant bound for Australia. A diary which he kept during the voyage shows that he was adventurous and interested in new places, that he had had little schooling but had read widely, and that he was somewhat pompous and self-opinionated. The *Triton* reached Hobart on 20 January 1824, and Robinson decided to disembark there. During the next five years he engaged in his trade of builder and contractor and made a lucrative business of it, for the colony had need of artisans.

Robinson was by nature serious, and by religion a Methodist. This combination led him to take part in religious activities of a missionary kind and in social activities directed towards the betterment of those not so fortunate as himself. In his missionary work he took religion to prisoners in the gaol and to seamen on the ships. He was also one of the founders of the Hobart Town Mechanics' Institute, and its first chairman, though he held that position for only six months, a quarrel with other members leading to his resignation.

This was his life until the beginning of 1829, when he threw up everything to live among the Tasmanian Aborigines and make friends with them in order to save their lives. Robinson tells us in the draft he made for the introduction to a book about the Aborigines, a book which he hardly started to write, that—

My mind had become early and deeply impressed with the deplorable state and condition of the aboriginal inhabitants. I was anxious to know if they could be instructed and whether anything could be done for their moral, religious and material improvement. Everything was said by a certain class to their prejudice. They were represented as a bare remove from the brutes and incapable of instruction. I could not believe that man could be so debased. Even brutes were accessible to kindness; how much more man, made after the express image of God. I felt persuaded that by kindness these poor creatures could be brought to a sense of their obligations and be made useful members of society.

Robinson's opportunity came early in March 1829, when there appeared a government advertisment seeking 'a steady person of good character' to reside upon Bruny Island and effect an intercourse with the Aborigines in order to ameliorate their condition. Robinson applied and was appointed. The next five years were to show that Robinson had the combination of qualities which made possible an intercourse with the race and the amelioration of its condition. First and foremost, he believed that the Aborigines were part of God's creation and therefore had similar rights to other men. He also believed that their degraded condition was due to their lack of opportunity to improve it and not to any lack of a capacity for improvement, that kind treatment and a cessation of persecution would lead them to give up their attacks upon the settlers, and above all he believed that he was the man to save the Aborigines, from themselves, as well as from others.

It all began at Bruny Island and against a background of ill-treatment, cruelty and attacks upon the Aborigines by the settlers and an effective retaliation by the natives. The acts of ill-treatment were the work of the stock-keepers, bushrangers, soldiery and sealers, but the population as a whole condoned the actions of their servants because they wanted to be rid of the natives.

Robinson spent nine months at Bruny Island getting to know the natives there, learning their language, making friends with them and gaining their confidence. During that time he came to realise that an important cause of the clash between the Aborigines and the settlers was the activities of the sealers who colonised a number of islands in Bass Strait. These men obtained native women, whom they held in slavery, by raiding the coastal tribes. Quite apart from the conditions under which the sealers kept their women and the resentment which the natives held against them for removing their wives and daughters, the effects on tribal economy and birth rate were considerable. Some idea of the magnitude of the effects on the lives of the natives can be gained from a list which Robinson made in November 1830 of those then alive in north-eastern Tasmania: there were seventy-two men but only three women.

Robinson's first direct information about conditions in the Strait came from two native women who had been brought to Hobart in October 1829 and sent to the Bruny establishment. These women had lived with the sealers and they gave him an account of the conditions under which they lived, the hard labour demanded of them by their masters, the punishments inflicted on them and their hardships. At this time Robinson was planning his first expedition — he wished to conciliate the south-western tribes whose aggressions against the colonists were causing anxiety — and so, after spending five-and-a-half months visiting those tribes, he decided to find out what was actually happening in the Straits.

Robinson arrived at Woolnorth, the Van Diemen's Land Company's settlement at the north-western extremity of Tasmania, on 14 June 1830, and from that time he was to learn much about the sealers, first those in the neighbourhood of Robbins and the Hunter Islands, and later those of the Furneaux Islands. At the time Robinson came in contact with the sealers, the sealing industry had come close to extinction as a result of the unrestricted killing of the seals. Many of the original sealers were dead or had left the islands, and there remained a relatively small number, many of whom had settled down to a subsistence existence on the islands, killing a few seals when they could find them, taking the mutton bird for its feathers and the kangaroo for its skin, and selling these things for the basic necessities of life, to which many added the produce of the gardens they cultivated and the pigs they raised. The key to the sealers' economy was the native women they held in slavery, who not only attended to the personal wants of their masters but were the work force in their foraging from the natural environment.

It quickly became apparent to Robinson that the sealers must be controlled and their native women taken from them and returned to their husbands and parents. Because his plans included the use of the islands for a place on which to settle the Aborigines he intended to remove from their Tasmanian homeland, he also had it in mind to evict the sealers from any part of the islands which he required for his own purposes. It was natural that the sealers, who for so long had made their own laws, should resist in every way they could the imposition of controls, their eviction from their homes and, especially, the taking of their women from them.

Robinson had a fair measure of success in all his endeavours, but he encountered many difficulties, and was only partly successful in removing the native women, both because the Government would not support him fully against the sealers and because many of the women were reluctant to go to Robinson's Aboriginal Settlement. This reluctance had its origin in part in the sealers' propaganda against the settlement and in part because the women did not wish to give up the 'luxuries' which the sealers gave them, as well as a certain freedom of movement.

Robinson spent a year in north-eastern Tasmania after his first expedition, from October 1830 to October 1831, and during that time made several visits to the islands in the Straits, saw to the formation of the Aboriginal Settlement, and made several expeditions by land in search of the natives inhabiting the north-eastern part of Tasmania. During his visits to the islands he learnt a great deal about the sealers. He made trips to many of the Furneaux Islands, on one occasion going as far north as Kent's Group, from where he had a fine view of the coast of Victoria.

The beginnings of the Aboriginal Settlement in Bass Strait were made at Swan Island on November 4 1830, when a party of natives whom Robinson had met a few days earlier in the country of the headwaters of Ansons River on the east coast of Tasmania and had persuaded to accompany him, were taken there. Swan Island was little more than a temporary holding camp. Over the next two months other natives came to Swan Island, so that by the end of December there were thirty-four on the island. Swan Island was clearly unsuited as a place of permanent settlement and so, on March 16, 1831, it was abandoned and the natives were taken to Gun Carriage (Vansittart) Island, on the way there spending a few days in temporary quarters on Preservation Island. The occupation of Gun Carriage by the natives meant the eviction of the sealers living there.

The Aboriginal Settlement remained on Gun Carriage until the middle of November 1831, that is, about eight months. When the people were first moved there, Archibald Maclachlan was put in charge of them. He was a convict and, before Robinson had brought him to the Straits at the time of the move from

Swan Island, he had been medical dispenser at the convict establishment at Maria Island. As the person in charge of the settlement, Maclachlan had the disadvantage of being a convict; he was also a weak man, so that he fell under the bad influence of the sealers.

Maclachlan was replaced by Sergeant Alexander Wight of the 63rd Regiment on July 5, 1831, Maclachlan remaining at the settlement as medical man. Wight's military training and the inflexibility of his non-commissioned rank led him to run the settlement as a concentration camp: in fairness it must be said that he had none of the qualities which could have led him to treat the Aborigines with caring kindness.

In the middle of November 1831, the settlement at Gun Carriage was abandoned and the Aborigines were moved to Flinders Island, a settlement being formed for them on the west coast of the island opposite Green Island, at a place called The Lagoons. It was not long before this situation was found to be quite unsuitable for a settlement, located as it was on what was little more than a sandbank between the sea and an extensive area of marshy ground which was flooded for much of the year.

Wight remained in charge of the Aboriginal Settlement until the beginning of March 1832, when he was replaced by William James Darling, an ensign of the 63rd Regiment. Darling was a young and energetic man, and from the beginning of his association with the Aborigines he showed them that he had their interests at heart and was sympathetic to them. Governor Arthur had formed the Aboriginal Settlement as a place to which the Aborigines of mainland Tasmania could be moved so as to prevent clashes between them and the settlers, but he also wanted to 'civilise them', that is, to change their way of life to a European form and to make Christians of them. It was Darling who brought about the first stages of these changes.

Within a week of Darling's arrival the settlement had become calm, and he had embarked upon a course of fair and sympathetic treatment of the Aborigines which induced in them a reciprocity of good feeling, a situation made clear in the accounts of the visits of James Backhouse and GW Walker. The instructions, which Darling received from Governor Arthur before he went to Flinders Island, were not only to keep the natives secure but also to civilise them. Arthur's concept of the processes of civilisation was to change a savage people into a community of God-fearing black Englishmen. Darling set out to interpret Arthur's policy of conversion as one which made the Aborigines self-supporting in the European style. He had to begin almost from scratch, by teaching the natives the basic activities of that way of life, even such simple procedures as the use of household utensils, but he soon succeeded in having them adopt European forms of living. Darling's plan for converting the people into a European-style community was to make farmers of them, and so give them independence. Unfortunately, Arthur did not give him the support he needed to do this, and so Darling did not succeed in effecting that change, although it seems that the Aborigines were then sympathetic to the idea. As a result there was lost a means of obtaining a meaningful future for the Aborigines in a black-white partnership, a form of association which could have succeeded under the conditions of the period, when peasant farming was still a way of life for a community.

Darling soon realised that the site at The Lagoons could not provide what was needed for an Aboriginal Settlement, that is, a place where the natives could have individual houses and gardens. He therefore set about looking for a more suitable place and, as a result of a survey made by George Woodward, decided that land at what was then known as Pea Jacket Point, and later as Settlement Point, would be suitable. After some months of preparation, the settlement was moved there at the end of January 1833. Darling named the new settlement 'Wybalenna', meaning in one of the languages of the Aborigines 'black man's home'. This was to be the home of the Aborigines until the middle of October 1847, when they were transferred to Oyster Cove on D'Entrecasteaux Channel.

The first years of the occupation of 'Wybalenna' were those of expansion, even better facilities being provided for the Aborigines, culminating in the L-shaped group of brick houses built for them by Robinson in 1837. Unfortunately for the natives, their friend WJ Darling remained with them only until about the end of July 1834, when he left to go with his regiment to India.

Darling was succeeded by Henry Nickolls, an appointment which must be regarded as a stopgap pending the arrival of GA Robinson, who earlier had been promised the position of Commandant at the settlement when he had brought in all the Aborigines living in Tasmania. Nickolls tried to give effect to Darling's idea that the Aborigines should become farmers, but he was not only frustrated by Arthur in this but clashed with him over the importance to be given to the indoctrination of the natives with what was then thought of as Christianity.

Nickolls was replaced by GA Robinson in October 1835.

*Built by the hands of Tasmanian Aboriginals, this restored chapel, once a shearing shed, is at Wybalenna, Flinders Island. It is at the same time a memorial to the alien culture imposed on the last of the Tasmanian Aboriginals and the only monument they have left behind them.*

*Bass Strait in the old days. Here a typical, double-ended island boat is taking visitors from Green Island to Flinders Island in 1893.*

*Cemetery at Wybalenna, Flinders Island, where the bodies of most of the Tasmanian Aboriginals deported by George Augustus Robinson in the 1830s are buried.*

Robinson remained at 'Wybalenna' until February 1839, when he left to take up the protectorship of the Aborigines at Port Phillip. Robinson did a great deal to improve the lot of the Tasmanians at 'Wybalenna', not only providing them with good housing but introducing a system of native police by which they had some influence over their own affairs. Robinson also introduced the natives to the European system of buying and selling. However, Robinson was basically a rather stupid man and he had little idea how to solve the 'Aboriginal problem'. During his stay at 'Wybalenna' he had really only one thing in view, to show that the Aborigines could be good black Englishmen; but he did not think out the consequences of making them so. There was the silliness of the Romantic names with which he dubbed the natives, but Robinson's real stupidity was to think that catechetical religion was real religion, and he and his helpers literally stuffed them with it, an undigested mass which was soon voided.

While Arthur was Governor, there was a real official effort to help the Aborigines, but with his departure the only concern was to spend as little as possible upon them. Franklin initiated this new regime, and parsimony was the prime mover in all matters connected with the Aborigines thereafter, only Denison of those who followed Franklin giving any real thought to their care. Progress at 'Wybalenna' therefore ceased on Robinson's departure. He was succeeded by Malcolm Laing Smith (1839-1841), Peter Fisher (1841-1842), Henry Jeanneret and Joseph Milligan, one or other of these two last being in charge between 1842 and 1847, interchanging in a Box and Cox fashion. After Robinson left, the Aborigines became no more than pensioners of the State.

Nothing has been said above about the Aborigines themselves, for whom the Aboriginal Settlement was set up. Their voice is seldom heard amongst the sound and fury of what is essentially a record of events under the control of and carried out by their usurpers, the settlers of Van Diemen's Land. They went to the settlement in the hope that they had some future, but the stream of deaths among them — from diseases of European origin! — must soon have convinced them that their only future was death.

**Further reading:**
NJB Plomley, *Friendly mission: the Tasmanian journals and papers of George Augustus Robinson, 1829-1834.* THRA, Hobart, 1966
NJB Plomley, *Weep in Silence.* Blubberhead Press, 1987

# 5
# LAND-USE
# ON THE BASS STRAIT ISLANDS

## A brief history of agricultural change

### Robin J Pryor

More than 180 years have passed since Furneaux, Flinders and Bass investigated the complex of islands at the eastern entrance to Bass Strait, and men like Reed and Murray explored the western entrance. In one sense, the years have done little to improve the barren appearance and treacherous approaches of these islands, which early navigators catalogued in great detail. As we shall see, however, some radical changes *have* taken place in the island landscapes, and in the economic pursuits of the men who have lived there.

**Physical environment**

King and Flinders Islands, the main ones which we shall be discussing, are located almost symmetrically in Bass Strait, off the north-west and north-east corners of Tasmania. They are in the same latitude, about 40°S, and of roughly similar size, Flinders being 1374 square kilometres and King 1099 square kilometres; Cape Barren Island, with Flinders Island a part of the Furneaux Group, is 445 square kilometres in area. They both have average temperatures around 56°F, and a mean annual rainfall between 711-813 mm with a winter maximum. Strong westerly winds are common, but frosts are rare. They have some physiographic similarities too, although geologically the base rock of King Island is much older (Cambro-Ordovician sediments) than that of Flinders Island (Siluro-Devonian granites). Coastal plains, calcareous dunes, swamps and lagoons, and small outcrops of basalt are found on both islands, and Mt Stanley in the plateau country of King Island rises to 183 metres; however, the granitic Strzelecki Peaks on Flinders Island rise to 762 metres, and Mt Munro on Cape Barren Island reaches 716 metres. These differences in base rock and altitude are responsible for the presence of skeletal and residual soils on the mountain and foothill areas of Flinders and Cape Barren Islands. Nevertheless, podsols and podzolic soils predominate on all the islands, mainly as sands and sandy loams, some acid, some calcareous, and some highly organic in swampy localities.

This, briefly, is the physical setting. We will now return to the beginning of the nineteenth century and trace through the various phases of settlement and land development.

## 1799-1830

Captain Bishop of the *Nautilus* established the first settlement south of Sydney when, in 1799, he commenced sealing operations from Kent Bay on the southern side of Cape Barren Island. After five months' work he could send back 9000 prime seal skins. Within three years there were over 200 sealers and some whalers in Bass Strait. A small settlement was established on King Island in 1803, although the American Amasa Delano left for New Zealand water soon afterwards because wholesale slaughter had so drastically reduced the numbers of seals. Sealing was probably at a peak before 1810, but it was still the locating factor for settlement, and the source of livelihood for some until the 1830s and early 1840s.

It is of interest that most of the sealers lived on the smaller islands, like Gun Carriage (or Vansittart), Preservation, Woody, Badger, Big Dog, Hunter and Cape Barren. This was partly because of known safe anchorages and less formidable shorelines and coastal vegetation. They ranged from the 'western straitsmen' on Kangaroo Island, and David Howie on King Island, to the 'eastern straitsmen' of the Furneaux Group. James Munro on Preservation Island, one of the later and more colourful of the pioneers, styled himself 'King of the Sealers' and 'Governor of the Straits'.

John Oxley gave little encouragement to more diversified development on King Island. About 1810 he reported that there was 'no timber of any consequence', poor coastal soils and, although a considerable quantity of oil could still be obtained from sea elephants, seal numbers were rapidly dwindling. GW Barnard, in a survey of resources in 1827 for Lieutenant-Governor Arthur, concluded the sheep would be unable to subsist and that King Island was more suited to penal than to free settlement (although, about the same time, Dumaresq reported good soils, water and timber!). The passing of the sealing era was verified in the same year, in Arthur's report that 'many of the rocks and islands that once afforded a rich annual harvest are now entirely deserted'. To some extent mutton-birding supplanted sealing, because by the early 1830s over 750 000 birds were being taken each year (cf the 466 208 taken in the 1968 seaon and 162 000 in 1983).

## 1831-1880

The first break in the generally parallel histories of the main islands came in 1831. In that year the Flinders Island settlement for the remnant of the Tasmanian Aborigines was established. In August 1838, the first commandant of the settlement at Wybalenna, George Augustus Robinson, reported a total population of 159. These included eighty-six Aborigines, five military officers and their families, thirty-five convicts, and twenty-one free settlers. Twenty-three acres (9.3 hectares) of land were under cultivation, and sixty-two cattle, 1300 sheep, thirty pigs and fifty goats provided food for the short-lived and unsuccessful experiment in race-preservation. The surviving forty-four Aborigines were transferred to Oyster Cove in 1847 when it became clear that the settlement could not achieve its purpose. Only eighteen sheep, 150 cattle (fifty-six of these 'in the bush'), and eight working oxen remained on the island when it was leased to Captain Malcolm Smith for ten years at £30 a year.

Smith had lived on both King and Flinders Islands since 1836, and had been commandant at Wybalenna for three years after Robinson. He leased Flinders Island from about 1850 until the end of 1861, when MacLaine took up fourteen years' rental. In 1849, Smith had offered to care for Aborigines on King Island for £20 each per year; permission for this was refused, but in 1850 he was granted a year-by-year occupation of King Island, and also of islands adjacent to Flinders Island. About this time, David Howie was constable of Bass Strait islands, and he was employing half-caste women to hunt kangaroos and wallabies for him. King Island was first leased in 1855 (by R and P Turnbull of Melbourne) for seven years at £300 a year. Grazing, indeed any kind of land use on these islands, was very much of a subsistence nature at this stage, and disaster was certain when any additional problem arose. This was the case with the tare, Darling pea (*Swansonia lassertifolia*), which caused fatal blindness and madness in all stock on the coastal dunes of King Island in the late 1850s. As a result, hunters on sub-leases replaced graziers on the island. Even in the Furneaux Group, which did not experience this particular difficulty, hunters were probably more important than sheep and cattle grazing, which maintained only a tenuous hold. However, King Island began to witness other development about this time. In 1859 a telegraph line was constructed across the island; the Wickham lighthouse was built in 1861; gold prospecting commenced; a length of tramway was built to facilitate timber extraction, and one schooner-load of timber was even sent to England.

By 1869, when the last lease was taken up, the tare was declining and some sheep could be brought onto the island for fattening and early re-shipment. Encouraged by some success, grazing expanded on the calcareous sand hills of the west coast, on pastures of melilot (*melilotus indica*), spear grass (*Bromus maximus*), and tussock grass (*Poa spp.*). It was at this stage that a second problem presented itself: the stock disease 'coastiness' led to the abandonment of grazing around Surprise Bay and Yellow Rock River. Thirty years were to pass before it was accidentally discovered that the transfer of stock inland was an adequate cure for the disease.

### 1881-1916

In 1881, 6000 acres (2428 hectares) were reserved on Cape Barren Island for the so-called 'half-castes', the hybrid descendants of the twelve European sealers and nine Tasmanian and Victorian Aboriginal women who had inhabited the smaller island since the 1820s. In 1887 land was first alienated on Green, Badger, Cape Barren and King Islands, though not on Flinders Island until the early 1890s. In 1886 and 1887 the surveyor John W Brown investigated the timber and agricultural potential of Flinders, King and adjacent islands. In reference to Flinders Island he concluded that:

> Should the right class of colonists be induced to reside on favourable terms on this island they could hardly fail to establish an enterprising settlement . . . But the absence of regular communication to enable the producer to make certain of a market, so far as agricultural products are concerned, is a necessity which no doubt may be fairly left to private enterprise as settlement progresses.

He was certainly prophetic on the first point, but perhaps rather optimistic on the second.

Despite Brown's encouragement, settlement was still on a small scale, and development slow for a number of years. In 1887 there were only two families on King Island apart from the lighthouse-keepers; in 1892 a timber company set up to supply wood blocks for London streets, collapsed after only four months. Between 1901 and 1911 the population rose from 242 to 778, and during this decade the King Island Dairy Factory commenced operations, tin and gold leases were taken up, scheelite was discovered at Grassy, and fat stock production was increasing. In 1909 there were 2134 fat cattle, 2312 fat sheep, and 58 pigs on the island. Less development occurred in the Furneaux Group, where Cape Barren Island was the centre of the pastoral district. By 1896 there was a state school, post office, and Anglican church serving the Cape Barren settlement of about 100 people, with some nineteen graziers, nine farmers and four mariners on Flinders Island. About 1903 development was initiated in the south-east of Flinders Island, and in 1911 both Flinders and King Islands shared in a boom in land speculation. On King, ninety-two per cent of the land was taken up during the boom, though poorer areas reverted to the Crown fairly rapidly. On Flinders Island, speculation was also widespread, if less dramatic.

### 1917-1947

Soon after the First World War, a level of development was initiated on King Island which was not to be paralleled on Flinders Island for some thirty years. The development was on two bases: the formation of the King Island Scheelite Company in 1917, and a Soldier Settlement Scheme which commenced in 1919, when 2300 acres (930 hectares) were allocated in sixty-seven lots (another 10 000 acres or 4047 hectares were acquired in 1920).

Mining declined from 1921, when the market was glutted. Eventually, a new company was formed in 1938, and, with Commonwealth Government assistance during the Second World War, it has continued to the present day. Dairying gradually became more important than the fattening of cattle and sheep, and in 1927 there were about 15 000 fat cattle, 3000 fat sheep, and 1500 pigs on the island; the human population was about 1200. In the 1930s, while butter and pig production increased, there was very little cultivation, and bracken (*Pteridium spp.*), inadequate drainage and sand blows were causing some problems. The Agricultural Bank of Tasmania took over the land settlement schemes in 1937.

During this period, settlement on Flinders Island continued to expand only very slowly, and fencing, clearing and actual utilisation lagged behind selection. In particular, this was due to the capital cost of clearing and draining sufficiently large tracts. There was a brief burst of activity in the Furneaux Group around 1921 because of the demand for yacca gum for the manufacture of shellac (from *Xanthorrhoea australis*). And during the 1930s the Nelson's Lagoon and Bootjack areas of Flinders Island were drained to extend the grazing acreage.

### 1948-1970

In 1945, under the War Service Loans Agreement, the way was paved for a second phase of soldier settlement schemes, and in

1948 a new scheme was commenced on King Island. Some 193 200-acre dairy settlement blocks were established. This was one of the earliest schemes started, and it later became necessary to raise the optimum stock carrying capacity on both fat lamb and dairy properties (to 1200 fat lamb ewe equivalents, and fifty-five dairy cow and replacements, respectively) to allow settlers to establish a stable economic position; this was done by further development and/or increased acreages on some properties. There was still some difficulty in ensuring viability in the late 1960s when thirty-two properties were passed in for a combination of personal and economic reasons. Between 1955 and 1968, the island's total number of cattle holdings increased from 154 to 219, and sheep holdings from 42 to 108. In June 1968, 64 dairying and 47 sheep properties were in Agricultural Bank estates. During this period dairy cattle remained stable around 12 000, beef cattle trebled to 21 000, and sheep increased over eight times to 145 000. While butter production was still increasing (1967-68: 880 tons), there was a swing to the earlier emphasis on the fattening of cattle and sheep.

In 1947, in response to considerable agitation from Flinders Island, a Joint Committee of the Tasmanian Houses of Parliament was appointed to examine the island's potential for War Service land settlement. The committee concluded that Flinders Island was 'admirably suited' for a settlement scheme, and immediate action was recommended. Actually, the development of about 70 000 acres (28 330 hectares) in five estates did not commence until 1952. Because of the King Island experience, an adequate carrying capacity was adopted at the outset, and the abandonment of uneconomic units has been minimal. Some 80 soldier settlement blocks were established, averaging about 700 acres each. Between 1957 and 1968, the island's total cattle holdings increased from 121 to 142, and sheep holdings from 94 to 144. In June 1968, the Agricultural Bank estates had seventy-three combined fat lamb and beef properties, five dairy and two fat lamb farms. During the ten years to 1970 dairy cattle and pig numbers declined significantly, beef cattle increased slowly, and the number of sheep grew over five times to 200 000. Butter production was much less important than on King Island (1967-68: 60 tons)

### 1971-1984

Trends in land use and agriculture in recent years reflect Tasmanian and national changes, as well as more localised problems and prospects. Flinders Island, with a population of

1100 in 1983, experienced a small growth of about 100 people compared with 1976. Over the past fifteen or so years, sheep raising had become firmly established as the main type of farming, especially for wool production, with Corriedales and Polwarths predominating; early prime lambs are being raised on the western coastal plain, and shipped or flown to Tasmania and Victoria. From 1976 to 1983 beef cattle numbers fell from 36 500 to 20 000, and there are now ninety-nine cattle holdings; over the same period, sheep numbers rose from 151 000 to 228 000, and there are now 100 sheep holdings (with about five per cent of Tasmania's total sheep numbers). The decline in cattle and increase in sheep is due to a number of factors, including returns from sheep compared with cattle, freight problems, and the large turnover in property ownership. In the past six years about half of the properties have changed hands, some more than once, and new owners cannot afford the risk in returns from cattle, the long wait from the purchase of a cow to the sale of a calf, nor can they afford the capital to buy cattle. Lease-purchase of breeding stock has recently been introduced to overcome the last problem, but until recently sheep were cheaper to buy than cattle per unit stocking rate. Amalgamation and change in Agricultural Bank farms prohibits useful comparisons between Bank and non-Bank properties. Changes in property ownership have had diverse results—amalgamation, new owner-ship, and division among several owners. The trend away from cattle was marked by the closure of the butter factory in the mid-1970s, and no cash crop has been found to supplant grazing.

The 120 rural establishments, covering 74 600 hectares, had about fifty per cent of their area under pasture. The most recent available statistics (1982-83) indicate exports of 9345 sheep (62 per cent to Tasmania), 48 997 cattle (approximately equally to Victoria and Tasmania), and 6371 bales of wool (up from 4400 bales six years before, and 98 per cent to Tasmania); 162 000 mutton birds were taken in the season (down from 239 000 in 1968) and this industry is likely to continue to decline. Agricultural problems have included the noxious weed spiny emex which was controlled by chemicals on Flinders Island 1982-83 as weevils proved ineffective; the sheep ked disease was 'intractable' on Cape Barren Island, where the sheep continue in semi-feral conditions; damage by wallabies was particularly serious on Flinders Island and many were poisoned 1982-83, and wingless grass-hoppers were also a problem on the island, as on the east coast of Tasmania.

The population of King Island has declined slightly in recent years, from 2760 in 1976 to 2680 in 1983. King Island has a good rainfall and is ideally suited to pasture production, especially for beef cattle and dairying, but also for sheep, with Corriedales and Polwarths on improved pastures. Sheep raising is mainly for wool, but there are also prime lambs processed at the local abattoirs or shipped to Melbourne. Since the late 1960s cattle holdings have risen slightly to 230, and sheep holdings have fallen slightly to ninety. The 3000 dairy cattle in March 1982 were about one quarter of the number in 1968, though by late 1984 the number had risen to about 4,000. Beef cattle numbered 52 268 in 1982, a 150 per cent increase over the number in 1968. The 86 000 sheep in 1982 show a significant decline from the 145 000 in 1968, but wool exports, around 3 400 bales in 1981-82, are on the increase. Butter production stopped in 1979, and King Island Dairy Products now concentrates on butter fat (213 091 kg in 1984), which is used totally on the island for processed cheeses and other fresh dairy products; while the factory is expanding and diversifying, the numbers of producers are remaining stable at about twenty-eight, due to the problem of oversupply. The abbattoirs are now private enterprise, and King Island Exports turnout about 25 000 animals a year, about half of this through the abattoirs. Other recent trends include understocking in cattle due to the large numbers sold off the island; a swing to finer wool sheep and away from fat lambs because of freight costs; a degree of new land clearance for private owners, partly as a tax-hedging device; the limited commencement of feed lot farming with rye, corn, barley and wheat; and younger farmers are taking up properties, with in-migration especially from Victoria, South and Western Australia. With the severe cutback in mining of tungsten, molybdenum and scheelite, part-time work for farmers has been well nigh eliminated.

Recent agricultural problems on King Island have included the noxious weeds, ragwort, managed by grazing, and serrated tussock; some increase in sheep footrot, though this was generally contained by the 1983 drought; low blood selenium levels in several flocks, with trials proceeding with slow-release pellets; and a significant incidence of leptospirosis in stock from known infected herds at Circular Head.

### Conclusion

Land suited to agriculture is still available for development on King Island, Flinders Island, and a large proportion of Cape Barren Island. Transport costs will continue to be major limiting factors, and strong conservation interests on the two main islands will not always be compatible with existing farming interests or government agricultural and other development policies. Holdings are being amalgamated, farming techniques modernised, capital accumulation is steady, and economically rational changes are being made towards wool production on Flinders Island, and towards beef cattle and fine wool (and to a lesser extent cheese processing) on King Island. Alongside these modern developments, traditional cattle raising survives on Prime Seal, Erith and the Hogan Islands, including the practice of swimming animals on to and off these islands. Agriculture is inevitable closely linked with shipping, however, on Flinders and King Islands, and the government-owned service to the latter is both more reliable and more expensive than the largely private servicing of the former. Fish products continue as significant minor exports from both islands, and though they are not the focus of this paper, the problems of competition and industry structure are significant for the islands' economies.

In closing, it is also worth noting the absence of any integrated development or management policy for the Bass Strait islands as a whole. Agriculture has received considerable capital and scientific support over the years, especially in the context of soldier settlement schemes. The time is now ripe for a much more comprehensive effort to integrate, with due consultation with residents and community organisations, future developments of agriculture, fisheries, tourism, and recreational uses — both land and marine. The problem, as elsewhere, will be to integrate legitimate economic/commercial interests with the unique attractions of an historic, little populated, but attractive island environment.

### Acknowledgements:

The assistance of the following is gratefully acknowledged: The Principal Archivist, the State Library of Tasmania; and Tasmanian officers of the following bodies — the Agricultural Bank, Department of Agriculture, Forestry Commission, and the Australian Bureau of Statistics.

### Further reading:

JW Brown, 1887, 'Report on Flinders Island', 'Report on King's Island', *Tasmania, Journals and Papers of Parliament*, Volume 12

*Crown Lands Guide*, 1885, 1912, Government Printer, Hobart

GM Dimmock, 1957, 'The Soils of Flinders Island, Tasmania', CSIRO *Soils and Land Use Series*, No 23

AL Meston, 1947, 'The Halfcastes of the Furneaux Group' *Records, Queen Victoria Museum*, II, I, pp 47-52

TH Monger, 1928, *King Island*, 'King Island News' Print

S Murray-Smith, 1973, 'Beyond the pale: The islander community of Bass Strait in the nineteenth century', *Papers and Proceedings*, Tasmanian Historical Research Association, vol 20, no 4, pp 167-200

RJ Pryor, 1967, 'Problems of Rural Land Development, Flinders Island, Tasmania', *The Australian Geographer*, 10, 3, March, pp 188-196

PA Smith, *Moonbird People*, 1965

CG Stephens and JS Hosking, 1932 'A Soil Survey of King Island' CSIRO *Bulletin* No. 70

NB Tindale, 1953, 'Growth of a People: Formation and Development of a Hybrid Aboriginal and White Stock on the Islands of Bass Strait, Tasmania, 1815-1949', *Records, Queen Victoria Museum*, New Series, No 2, June

# 6
# MUTTON-BIRDING

## DL Serventy

Mutton-birding, or the taking for commercial purposes of the young of a sea-bird known as the Tasmanian Mutton-bird, or Short-tailed Shearwater, formerly known as one of the sooty petrels, is a peculiarly localised Australian industry. Tasmania is now one of the few places in the world where this sort of commercial wildfowling still survives, though it was once widespread in Europe. Up to the days of the English Civil War, in the 17th Century, the Earls of Derby used to receive as an annual rent a barrel of 500 salted petrels, of a species very similar to but slightly smaller than the Tasmanian bird. These were taken on an islet off the Isle of Man.

How the term 'mutton-bird' came to be used for these sea-birds, species of which occur all over the world, is still a subject for argument. It seems to be a word only used in Australia, but it has been current with us since the very earliest days of European settlement. The earliest documentary record was in 1790, but the name had been applied to another species at Norfolk Island. The general opinion is that the name arose because of the similarity in the taste of fresh birds to mutton, though some people cannot appreciate any resemblance. Others believe that the strong smell of the burrows is like that of sheep, and others again suggest that the very fat young birds, with their tallowy appearance, might have reminded the early colonists of mutton carcases.

The earliest exploiters of the Tasmanian species were the ship-wrecked crew of the ship *Sydney Cove* which was beached at Preservation Island, in the Furneaux Group, in 1797. Afterwards the sealers established on these Bass Strait islands subsisted on the birds. Ultimately an industry dependent on them became stabilised and unlike sealing and whaling, remained permanent and largely unimpaired despite the indiscriminate slaughter of the early years. Before explaining why the industry was able to endure into our own times, we must consider briefly the remarkable life cycle of the bird.

It is a strict migrant. It nests only on the islands of south-east Australia, from those off Ceduna in South Australia to Broughton Island off New South Wales, but Tasmanian islands are its principal strongholds. After the young are raised, the entire population, of many millions, departs for the North Pacific, where in the nutrient-rich waters around the Aleutians the birds feast on the planktonic life which in the northern summer also supports vast shoals of sockeye salmon, fur seals, whales and other marine creatures. At the end of September the returning adults make their first landfall on the nesting islands. They scratch out the old burrows, in which they will lay their eggs, and engage in an active and noisy 'courtship' behaviour. In early November the parents disappear from the islands for about three weeks until the egg-laying is due. This begins about November 21 and for the most part is completed by the end of the month. The peak of egg-laying is around November 25-26, and eighty-five per cent of the eggs are laid within three days on each side of this mean date.

The amazing thing is that this timing and synchrony have not varied at all since the first white colonists made their observations in the early part of the last century. Commercially, of course, this constancy is of enormous importance. It means that the fledglings which will be harvested later on in the season will always be of a marketable size at a set time of the year, and the catching time-table can be permanently fixed. Furthermore the highly synchronised egg-laying, over a very narrow period of days, means that all the chicks will be of approximately similar size and the operator need not pick and choose which bird is big enough to take—he *knows* that each time he puts his hand in the burrow to pull out a young bird it will be of a suitable size to put in his cask.

Incubation of the egg—for only one is laid each season, and not replaced if it is lost—averages 53 days. Hatching occurs between January 10 and 23, and the young bird remains in the burrow for an average period of about ninety-four days. The young leave for their northern flight from the third week in April to the first week in May. The eggs are brooded within the burrow in turn by the male and female, and these incubation shifts of duty are extraordinarily lengthy by usual bird standards. The male takes the first shift following the laying of the egg, and stays alone in the burrow, without taking any food, for about fourteen days; then it is relieved by the returning female. Similarly with the feeding of the chicks. At first the nestling is fed almost nightly, but after the first week there are gradually increasing intervals between meals, and towards the end of the fledgling period these may lengthen to as much as a fortnight or longer. During the final fortnight or so the parents abandon the young altogether, ceasing to feed them.

All the bird activity on the islands takes place at night. A remarkable feature about a mutton-bird rookery is the apparent desolation and utter silence of the day-time compared with the animation and clamour at night. The parents come in only after dusk, to change their duty shifts or to feed their young. They leave again at the first sign of dawn before the diurnal predators are stirring abroad — dangerous creatures like the big Pacific gulls, ravens and hawks. One may walk over the area in the day, and unless specifically searching for the birds or stumbling accidentally into the burrows, will have no inkling that many thousands of brooding mutton-birds or their young are underground — silent and unobtrusive.

The homing powers of the birds are extraordinary. Once a bird selects a plot of ground in which to excavate its burrow, or takes over an existing one, it may return to this spot permanently year after year, though the actual site may vary slightly. Young birds tend to return to the general area in which they are hatched. But they do not come ashore before they are three years of age, although they do frequent local seas after the first year's return migration. They spend a couple of summers on the surface of their home islands, learning their way around. Then, at five years, the females first begin breeding; and the males a year later, on the average. It has been computed that the average life expectancy, after attaining breeding age, is about fifteen years. Some individuals which were recorded as breeding birds in 1947 at the research island, Fisher Island in the Furneaux Group, were still breeding in 1975!

The huge food resources provided by the vast sea-bird colonies were tapped where accessible by the Tasmanian Aborigines, as the Maoris utilised a closely similar bird in New Zealand. With the decline of sealing in Bass Strait, the Tasmanian mutton-bird became the important item of subsistence for the resident sealers and their families. In the earliest days the exploitation was on a wholesale scale — adult birds, fledglings and eggs were taken indiscriminately. A very large trade was done in feathers and down for upholstery. There was a particularly large trade in mutton-bird fat in connection with the Tasmanian saw-milling industry. It was used in mills where logs or heavy flitches had to be drawn along wooden skids, and the liberal application of mutton-bird fat greatly facilitated movement. The fat was also used in greasing the skips in coalmines. For rendering into fat, whole birds were tried-out, and some of the boilers used still lie rusted in the rookeries. There is one now on Great Dog Island, south of Flinders Island. The harvesting was exceedingly wasteful and no thought was given to conservation.

As now stabilised, the industry is mainly concerned with salt-curing the young birds for food, with by-products in down, body fat and stomach oil. The industry is stringently controlled by the Tasmanian National Parks and Wildlife Service. The catching season, for the young birds only, starts on March 27 and ends on April 30. The adult birds and the eggs must not be interfered with. The Furneaux Group, in north-eastern Tasmania, always has been and still is the headquarters of the industry. Originally the sealers founded it, and their descendants, the Cape Barren Islanders, continued it, being joined in later years by newcomers to the islands. With the expansion of various farming and fishing activities in the islands it no longer holds pride of place as the mainstay of the Flinders Island economy, but it still is a picturesque and important annual event. In former times, everything revolved around the mutton-birding industry — the schools closed for the long vacation at the mutton-bird season, not for Christmas and the summer; farmers dried off their cows so they could go to the mutton-bird islands for the annual 'birding'; inter-island trading ketches plied between Lady Barron and the bird islands with people, supplies, mutton-bird casks and all the appurtenances of a thriving local industry.

Things have changed somewhat now. The school year now accords with that of the rest of Tasmania, and consequently families find it awkward to go birding as freely as they used to. A seasonal industry like this is also faced with the problem that adequate labour is scarce. It is difficult to find skilled operatives who can afford to leave permanent jobs to go 'birding' for a couple of months. The shipping situation, too, has altered, and transport to some of the outer islands, such as Babel Island, is expensive and unpredictable. So for these various social reasons the catching effort has declined and the annual output of processed birds has dropped. Whereas in 1933 the season's catch was about 800 000 birds, that of 1968 was just under 500 000. As will be shown later, this was not due to a shortage of birds but to a steadily declining catching effort. In 1985 the total commercial catch in Tasmania was 324 579 birds, 136 329 from the Furneaux Group in eastern Bass Strait, the rest from Three Hummock Island, Steep Island, Walker Island, Hunter Island and Trefoil Island in western Bass Strait. The Trefoil Island catch for 1985 of 83 210 birds would have been about 130 000 but for a freezer break-down, so it may be assumed that the present catching level is about 375 000. In addition there is an estimate of perhaps 300 000 birds taken on non-commercial licences.

The founders of the industry in the last century had developed an efficient 'factory line' process long before Henry Ford and other industrialists put it into general practice. On the islands the birding families operate a two-unit station located in spots convenient to boat-landings. One building serves as living quarters, a cosy hut with walls gaily decorated with pictures cut out of the illustrated papers—in recent years, coloured ones predominating. Even a simple thing like this was carefully thought out. The pictures were stuck on to the wooden walls with a paste made of flour and water to which sheep dip had been added to deter mice. The earthen floors of these huts were covered with tussock grass cut from the nearby rookery. Now, of course, these huts are much more modern, but somehow they have lost the warm character and charm of the old ones. The second shed is the processing unit, and its internal arrangements are ingenious. All are built to the same traditional pattern. The nameless originator of the system, whoever he was, surely deserves some recognition in our social history.

The birds are taken by hand by 'catchers', working the rookeries systematically. The young birds are killed by a dexterous breaking of the neck and threaded through the lower beak on to a wooden spit. The load of up to sixty birds is brought back to the shed and placed on a wooden rest at one end of the processing shed. The stomach oil is squeezed out into a collecting drum and the body then thrown through a window into a small room, the 'pluck-house'. Here the women of the party pluck off the down and pass the bird through a hessian-guarded flap-door into the 'scalding room'. The legs are severed, usually by the children, and the bodies scalded momentarily in a copper which is always on the boil. The remaining down feathers are rubbed off and the birds passed into a third room where the final processes take place. The bodies are cleaned and laid to cool on racks. When ready the birds are finally dressed and spread out flat like kippers, to be packed in casks with a liberal application of salt. Brine, made up with salt added to sea-water, and of a concentration to float a potato, is poured in. The residues are 'tried out' for fat.

The casked salted birds are sold in Tasmania and to a small extent in southern Victoria. Some are exported to New Zealand. In recent years an increasingly large number are packed in the fresh state, frozen, and transported from the islands by freighter planes. During World War II a fish cannery at Lady Barron, now disused, prepared a very appetising tinned product sold as 'squab in aspic'. The stomach oil is sold for pharmaceutical purposes, including a sun-burn lotion popular on Gold Coast beaches in Queensland; the body fat is rendered down for sale to dairy farms as an additive to skimmed milk for feeding to calves, and the down is used, among other purposes, as a filling for sleeping bags for bushwalkers. In 1945 the salted birds sold at the equivalent of $3 a hundred. By 1968 salted birds fetched $12 to $14 a hundred to the producer, and fresh birds $16. Oil fetched 75¢ a gallon (16.5¢ a litre). In 1985 prices for fresh and salted birds (the fresh birds, being frozen, costing more to transport and hence are slightly higher in price) ranged between $80 and $110 a hundred. Oil brought $2 a litre. Feathers, which in the post-war years sold at 4s 6d a pound (50¢ a kilo), have been subject to wild fluctuations, owing to limited demand, and at one stage in the early 1970s were fetching only about 10¢ a kilo. They now fetch $1 a kilo.

In recent years, with the conservation and control aspect in the air, the Tasmanian authorities have been increasingly anxious as to the general health, as it were, of the industry, wondering if the future was being imperilled by unwise exploitation and whether any other factors were affecting the welfare of the industry. The present investigations were set in train in 1947, and are being jointly conducted by the Commonwealth Scientific and Industrial Research Organization and the Tasmanian National Parks and Wildlife Service. To provide data for a population study, many young birds are banded at random over the commercial islands each March, immediately prior to the opening of the catching season. The number of banded birds recovered by the birders during the subsequent operations gives an estimate of the proportion taken in relation to the total young birds present. Because the birds nest in burrows, of arm's length or longer, and each bird has to be sought separately in burrows hidden among fairly dense tussocks, it is virtually impossible to take every bird present, as would be the case with a surface nester. These banding techniques show that the harvest on well-birded islands may be as low as fifty per cent of the young birds present, though at times it has reached about seventy per cent. These figures, together with the catch statistics, are supplemented by counts made in sample plots over the rookeries.

It has been found that in very productive regions, where the soil condition is favourable, there may be as many as 2500 occupied burrows an acre (6200 a hectare), as in portions of

Babel Island. However, the average productivity is much less. The soil may be too hard, or becoming too shallow. Birds cannot burrow where the soil has a depth of less than 9-10 inches or 25 cm. It has been found that on some islands the carrying capacity for mutton-birds has been very seriously impaired by the grazing activities which have taken place in previous years. Both sheep and cattle, apart from the obvious damage that is done by trampling down of burrows, tend to compact certain types of ground, making it difficult for the birds to burrow, and the burrows are spaced further apart. On some of the South Arm rookeries, south of Hobart, the turfed surface, too hard for the birds, is burrowed into by rabbits and these holes are then invaded by the mutton-birds. If an island is to be preserved for profitable mutton-birding, every endeavour should be made to retain it under natural conditions and human activity should be restricted to the minimum. Grazing and other forms of husbandry should be excluded. After all, very large parts of Australia are available for sheep and cattle grazing. The comparatively few islands in south-eastern Australia, having unique advantages for commercial mutton-birding, might well be left exclusively for that purpose.

# 7

# FISHING IN BASS STRAIT

## Alistair Gilmour

The history of fishing in Bass Strait is difficult to unravel from the stories of the development of Victoria and Tasmania.

Bass Strait is the greatest area of continental shelf within easy sailing distance of the main centres of population of the Australian continent. The average depth is of the order of 40 fathoms (65 metres) and it is about 130 miles (210 km) north to south and 340 miles (550 km) east to west, measured on the 39th parallel. The superficial similarity with the waters of the North and Irish Seas must have appeared strong to the early settlers, especially when they felt the need for better fish supplies.

An early industry that was prosecuted with vigour in Bass Strait was sealing, which commenced in the latter half of 1798 and had passed its zenith by 1806. However, when Dumont d'Urville sailed into Westernport, in 1826, through the western entrance, he came upon a party of sealers camped there, and his artist, Louis de Sainson, depicted the sealers them carrying their catch. Whaling was another of the early commercial interests in the latitudes of Bass Strait, but it, too, soon fell into decline. In 1828 the value of whale products exported from New South Wales was £27 000 and it reached a maximum of £197 000 in 1838. By 1894 the wholesale slaughter of the whale stocks, as with the seals, had reduced the value to a mere £700.

However, while the explorers, the whaling crews and the sealers carried on some fishing in the waters of Bass Strait in the period around the early 1800s, little is known of the actual fishing industry. As the Tamar estuary was one of the earliest settlements on the shores of Bass Strait, it is probable that a small localised fishery was developed from the early part of the century. The same was probably true of the Victorian coast, where it has been traditionally an inlet and coastal fishery since early settlement.

Various commissions and committees of enquiry into fishing and the fishing industry have been held in the States with access to Bass Strait. The Victorian report, published in November 1892, recommended *inter alia* that, as there appeared to be good trawling grounds off the coasts:

1) a careful survey be made of the sea bottom in Bass Strait;

2) the government should obtain trawling gear and lend it to selected fishermen; and

3) the government should fit out a steamer to collect the fish — what would be called today a mother-ship.

All were most concerned with the need to improve fish supplies and to develop the offshore fisheries. Other limitations were noted in the 1896 New South Wales Royal Commission and the earlier, 1892, Victorian Select Committee, where concern was expressed about the methods of transporting the fish from the point of landing to the markets, about the state of the markets themselves, that the fishermen complained of losses of their consignments to the market, and that pollution was causing the loss of profitable fishing grounds and areas considered nursery grounds.

It is of interest to note that inquiries in 1909 and 1960 had many of these same points to make.

A 1919 Victorian Royal Commission found it necessary to comment on 'A Neglected Asset' — the trawling of offshore fish. A Tasmanian Royal Commission three years earlier, conducted by Professor TT Flynn of the University of Tasmania, probably more widely known as the father of Errol Flynn, also had lamented the lack of interest in the 'Outer Grounds' or open sea. Professor Flynn noted that although steam trawlers were operating in New Zealand, 'It is the practice in Tasmania — as it is, in fact, almost all over Australia — to confine the fishing to the Home and Middle grounds'.

These early twentieth century inquiries had the benefit of the investigations and reports of HC Dannevig, who worked from the Commonwealth research trawler *Endeavour*. One of these reports on Australia's fisheries, issued in 1913 by the Department of Trade and Customs, showed that Dannevig, who was Commonwealth Director of Fisheries, had little faith in the potential of Bass Strait itself to produce paying quantities of demersal, or bottom living fish. Rather he recommended the east coast off New South Wales, extending along the eastern entrance to Bass Strait to about the latitude of the south end of Flinders Island. In addition he recommended the edge of the continental shelf in the Great Australian Bight as 'ascertained trawling grounds' between the longitudes of about 126° and 131¼° east. Unfortunately the vessel, crew and Director were lost on a return voyage from Macquarie Island in December 1914, and so too, it appears, was the determination to explore further the areas that Dannevig had not been able to evaluate to the full.

Nevertheless three steam trawlers, designed to use the standard Northern Hemisphere otter trawls, were brought out from England in 1915 by the New South Wales Government to fish from Sydney. Four further trawlers were built in Newcastle in 1920 to join the fleet, but by 1923 the Government had incurred such losses that they were sold to private firms and individuals, who in turn managed to make handsome profits. The trawlers were soon forced by poor catches, probably as a result of over-fishing, to fish further to the south, and had turned their attention to the area south of Cape Everard where large quantities of tiger flathead (*Neoplatycephalus sp.*) were found. These too were reduced to unprofitable levels, and in 1958 otter trawling for scale fish in south-east Australia ceased altogether.

An alternative to trawling was introduced in the 'thirties in the form of Danish seining — still in Australia popularly called trawling — where the method of fishing is to run out, in the form of a triangle, the hauling rope, the net along the base and the other rope up to the apex. The winch is then used to haul in the ropes and net to the boat until the catch is landed. This method allowed the exploitation, with smaller boats and crews, of the more restricted areas of sea bed not able to be fished by the larger steam trawlers. Overseas it is not a method of fishing which can compete with today's modern automated stern-trawlers, which are as small as forty-five feet (fourteen metres) in length, except on the particularly patchy ground unsuitable for trawling.

The grounds closer to the Victorian port of Lakes Entrance were exploited by the Danish seiners after the fishery off the East Coast became unprofitable in much the same way as the trawl fishery, and by 1947 some boats had moved into Bass Strait waters. Here the grounds appeared so limited that some of the boats moved on to San Remo within the year, and by 1965 there were only seventeen Danish seine boats licensed in Victoria. This method of fishing has not been developed to any extent in Tasmanian waters, but in fact it lands a reasonable percentage of the fresh fish catch from Victorian waters. It was suggested in a 1966 report that this fishery could be extended by using more of the species caught rather than discarding them at sea, and by the introduction of modern stern trawlers to fish the eastern slopes of the Strait.

In 1974 trawling recommenced in Bass Strait with operations on the continental shelf and slope off eastern Victoria, in northern Bass Strait and on the continental slope off Western Victoria. There has been a concurrent decline in the number of Danish seiners, following the conversion of most of the seiners working off the south-east of New South Wales to otter trawling. In Bass Strait the main species caught to the east are gemfish, jackass morwong, and redfish, in the northern area tiger flathead, jackass morwong, latchet, ocean perch, warehou, snoek and common sawshark, and in the west blue grenadier, king and silver dories, gemfish and ling. Danish seining is still carried out off Lakes Entrance and in northern Tasmanian in-shore waters for school whiting, tiger flathead and flounder. The trawl fishery is still only in its infancy in Tasmanian waters.

Major ports are Eden, Lakes Entrance, Port Albert, Port Welshpool, San Remo, Newhaven, Apollo Bay and Portland on the northern coast, while Stanley and Hobart service the Tasmanian coastal areas. The boats operating on the western side of the Strait move up into New South Wales in winter, and this fishery is the only one thought to have much development potential.

A new fishery in Bass Strait started when scallops were discovered off Lakes Entrance in 1970. Subsequent beds have been identified around the Furneaux Group and in western Bass Strait. Another fishery that has gained importance in Bass Strait since the early 1960s is the abalone (mutton-fish) fishery. Australian production contributes about one-third of the world production, and about one quarter of Australian production comes from Bass Strait.

Until recently the four most important fisheries associated with Bass Strait were those for the edible sharks, the Snoek (or Barracouta) (*Leionura atun*), the Southern Crayfish (*Jasus lalandei*) and the Australian Salmon (*Arripis trutta*). Records of the landings of Snoek have been kept since 1911, and there are figures for the export of the species from Hobart in the latter half of the 19th century. The fishery was based on the Bass Strait ports, and the majority of the catches were taken within twenty miles of the home port in 10-30 ft. (4-10 m) open boats using hand jigs. The fish is known to occur out to the edge of the continental shelf to the 28th parallel, and it is known from shores of all the southern continents between 30 and 45° south. Biological research work has been carried out on the species since the 1940s, but the problem that was present then is still the main source of concern—how to dispose of the product once it is caught, as ample supplies are landed each year. A trawler with a Russian whaling fleet, when it visited Melbourne in 1964, caught seven tons of barracouta in fifteen minutes while fishing three miles off Cape Schanck.

Open or half-decked wooden boats of six to ten metres still operate the fishery. Research workers suggested in the 1950s that catches from the autumn fishery were low due to changes in the behaviour of the fish, as a result of changes in the availability of its food organisms close to the coast, making the shoals of snoek unavailable to the small boat fishery. A decline occurred in the fishery after the war-time high, and a more recent drop in production has occurred since the early 1970s.

Production in recent years has been below average and Victorian fishermen have experienced a declining catch per unit of fishing effort. Increases in production costs in the recent period have exceeded price increases, and this may account for the decrease in production.

The Australian Salmon—a sea perch and not a true salmon—is also fished from shore. Beach seines are used to surround large schools spotted from the air. The present heavy fishery is a feature of the last thirty-odd years only, but it seems certain that the early settlers were familiar with it and may have confused it with the true salmon of the northern hemisphere. Research has been carried out on the species by CSIRO Division of Fisheries and Oceanography for some years. Current thinking is that the fishery can continue to prosper as long as present methods of capture do not change.

The edible shark fishery has traditionally formed the bulk of the landings from Bass Strait for Victoria and Tasmania. The former is based on two main species, the School shark (*Galeorhinus australis*) and the Gummy shark (*Mustelus antarcticus*), although a few other species are included in the landings. It is known that the School shark has been fished in Tasmania since 1880, when the liver-oil and fins were exported to China. This is truly a Bass Strait fishery as the fishing vessels, of 38 to 80 feet (13-30 m) in length, range over most of the area from ports in both states. The gear is mainly long-line with some gill netting, and since 1964 some of the boats at Lakes Entrance have turned from fishing for Skipjack Tuna (*Katsuwonus pelamis*), using monofilament gill net, to the better-paying shark. Fishermen complain that they must range further from their port in order to land good catches, but it has been argued that the stocks are underfished at present. Here is a situation that requires incisive study to elucidate this point. In 1966-67 landings were slightly up on previous years, as were those of 1965-66, but this may be due to the increased effort resulting from the use of monofilament gill nets.

Early in 1972 the Commonwealth government's analytical

laboratories (then part of Customs and Excise) rejected the import of a shipload of shark fillets from New Zealand, on the grounds that the mercury content of the flesh was in excess of allowable levels. The New Zealanders were quick to point out that the fish were probably part of the same stocks as those fished in Australian waters. This sparked an Australia-wide examination of mercury levels in fish, and the introduction in Victoria of a maximum size of shark beyond which fish could not to be sold to the public, as the older the fish the more mercury carried.

The southern cray fishery is based on what is called elsewhere in the world a spiny lobster, and belongs to a group distributed from South Africa to New Zealand. Other species are found in South America, the west coasts of North America and Europe, in the Mediterranean and throughout the tropics. The fishery has existed since last century, but did not develop to its present level until after the Second World War, when vessels started to range further afield. Now Victorian vessels fish throughout the islands of the Strait. There is currently a monitoring program, organised by CSIRO and the States, to obtain catch and effort statistics and information on lengths of the fish landed, in an attempt to manage the fisheries in order either to maintain production levels or to improve them. The cray is a major export earner, amounting to over $145 m, in 1984-85 ($23.5 m in 1867-68). In this the western cray is the major component. This resource is one which requires continual monitoring to enable rapid management decisions to be made in response to changing conditions in the fish stocks.

So far the discussion has been restricted to the fisheries that pay sound returns at present; what of the pelagic or surface and mid-water fishes that are major components of overseas fisheries? Apart from Snoek and Australian Salmon, the prospects in the Bass Strait lie with the extended exploitation of the tunas, the Jack Mackerel (*Trachurus novae-zelandiae*) and the clupeoids such as pilchard.

The pelagic fisheries of Australia have attracted attention from the earliest times, and many good resolutions have been passed in relation to this endeavour, as in the other fisheries. In 1880 a Royal Commission was appointed in New South Wales which stated, 'The Maray (Pilchard) appears annually in immense shoals about midwinter, passing in a northern direction'. A conference, held in Sydney in 1929 after a preliminary one in Melbourne, resolved to recommend to the Commonwealth Government a thorough investigation of the

74

pelagic fish resources of Australian waters. A purse seine vessel was commissioned but met with little success, and the net was lent to private fishing interests in Tasmania. Subsequently some success was attained with landings of approximately twenty tons on a number of occasions.

Anchovies temporarily came under some increased fishing pressure in 1946 as a Melbourne firm started to use them in fish paste; the boom did not last long! In 1960 another Melbourne firm began to can pilchards but that lasted for two years only. However, also in 1960 a purse seine fishery commenced off Lakes Entrance to supply a fish-meal plant established at that port.

Australians have talked for some time about the prospects for a fishery for Jack Mackeral, but as yet no successful enterprise has been established. Experimental pelagic fishing from Lakes Entrance and Triabuhna have not been over successful. It is interesting to note that Japanese interests are producing some 20 000 tonnes per annum from a demersal fishing operation in New Zealand waters.

Dannevig noted in his 1913 report that the Japanese had obtained their first English trawler six or seven years previously and had by that time built a further hundred, which were even then to be found on the China and Korean coasts as well as their own. While it is invidious to draw such comparisons, one cannot but feel that there may have been some reluctance on the part of fishermen in the past to develop new resources and on the part of the public to buy fish, at least for a trial, unfamiliar to them. The 1960 experiments by the Commonwealth Government with the otter trawler *Southern Endeavour* in the Great Australian Bight showed that, while it was possible to land fish in paying quantities, the public would not adapt to new species of fish.

One aspect of the development of the fisheries of Bass Strait in general, but particularly of the pelagic resources, that continues to be passed over, is the thorough assessment of the stocks and the ecology of the species while still unexploited. If the errors of the sealing and whaling industries are to be avoided, and a possible clash of interests prevented, it would seem advisable that planned development — which implies informed development — should be undertaken. There are major oil and gas reserves under the floor, there may be minerals to be recovered from both the surface of the sea bed and from the sands that form a major proportion of the substrate; in time minerals could possibly be extracted from the waters of the Strait, and it may be that the fishery that has been the source of so much speculation over the last two centuries may come about. With these possibilities in mind, one turns to search for information on the physico-chemical data that would enable evaluation of the area to be made; to the fund of ecological knowlege with which to manage the development of the resource: and both with the same result — little or nothing.

As a direct consequence of this lack of knowledge the potential — with respect to fisheries at least — of Bass Strait is unknown at this stage and cannot be known until detailed studies are carried out in the region. It is difficult to think of a similar shallow sea, close to major centres of population where modern technology is well developed, that is as little studied as that on Victoria's doorstep.

When writing the earlier version of this article, the concluding remarks noted the lack of research into the physical environment, the sorry state of our knowledge of the life-histories of the important species, and the general lack of research interest on the part of both universities and government. That has all been changed. Fisheries services have established the basis of a number of fisheries and their rational exploitation. Exploratory fishing by researchers, using both chartered and official vessels, have delineated some of the new fisheries. Researchers from marine science laboratories have worked on the biology of some of the important commercial species and on the chemistry of Bass Strait waters.

Most importantly, the universities, with the help of the Victorian Government, came together to form a marine science consortium, called the Victorian Institute of Marine Sciences, in 1977. The Institute soon moved to establish a Bass Strait Study, launched at a workshop in July 1978, which continues active research.

**Selected further reading:**

*Australian:*

Roughley, TC, Fish and Fisheries of Australia. Revised 1966. Angus & Robertson Ltd, Melbourne

Scott, TD, The Marine and Freshwater Fishes of South Australia. 1962. Government Printer, Adelaide

Burdon, TW, The Victorian Fishing Industry, 1966, Fisheries and Wildlife Department, Victoria

Anon, Australian Fisheries (published monthly). Fisheries Branch, Department of Primary Industry, Canberra

*Overseas:*

Borgstrom, G, Atlantic Ocean Fisheries, 1961
Heighway, AJ, Fishing News (Books) Ltd, London
Graham, M, Sea Fisheries and their investigation in the United Kingdom. 1956. Edward Arnold (Publishers) Ltd, London

# 8

# HAS KNOWLEDGE TAMED THE FRONTIER?

## Laurie Hammond

When these original talks were broadcast by ABC Radio in 1969, several of the speakers lamented the relative paucity of detailed scientific information about Bass Strait. It was suggested that among the remedies would be the development of concerted research efforts within the Victorian universities and the establishment of a marine laboratory for teaching and research purposes.

In the last fifteen years, these goals have been realised. Marine sciences are increasingly taught at educational institutions in Victoria and Tasmania, several marine laboratories exist on the shores adjacent to Bass Strait, and intensive research programs have commenced in universities and State and Federal government laboratories. The purpose of this article is to record the emergence of these institutions and research programs, to reflect upon their achievements, and to ponder whether Bass Strait, our 'last frontier', still lies beyond the horizons of our knowledge.

Undoubtedly, these advances were assisted by events occurring Australia-wide. Perhaps partly spurred on by a generally heightened environmental awareness and interest in the sea and its resources, there developed during the 1970s a general recognition in Australia that marine science and technology would be of crucial significance to the nation. It was during this time that the Australian Marine Science and Technology Advisory Committee (AMSTAC) was formed (providing advice initially to the Prime Minister, later to the Minister for Science) and the Australian Institute of Marine Science was established at Townsville.

In this favourable climate of strong political and community interest in the sea, Victoria developed two institutions that have done much to alleviate our ignorance of Bass Strait. The first of these arose out of several research groups formed in different government departments between 1968 and 1973. They began to work on a range of marine problems, mainly in inshore waters such as Port Phillip Bay and Westernport Bay, and focussed closely on pollution and environmental protection. In 1979 the units were amalgamated and became in turn the Marine Sciences Laboratories and then the Marine Resources Management Branch, located at Queenscliff. This is a large marine research centre by Australian standards, and essentially has retained its focus on fisheries and pollution problems in inshore waters. However the lab also has initiated a number of investigations in Bass Strait, principally into the fisheries resources such as scallops and the finfish caught by bottom-trawling in eastern Bass Strait.

At the same time, a group of far-sighted Victorians began planning and plotting a course towards an exciting goal: an institute of marine sciences for Victoria. Prominent among them was Dr PG Law, the Australian Antarctic expeditioner and scientist. He was strongly supported by Professor John Swan of Monash University, who was to become chairman of AMSTAC. For several years they proselytised and prodded the government, the research community and the universities. Their words fell on receptive ears for in 1974 the Victorian Institute of Marine Sciences Act was proclaimed, to be followed by further legislation in 1977. The Institute began operating in 1978, established as a statutory authority reporting through the Minister for (at present) Conservation, Forests and Lands.

There were many visions of the Victorian Institute of Marine Sciences, some grand, some more humble. However, they all contained an element which is now central to the *modus operandi* of VIMS: the Institute was to occupy a pivotal position in the marine science community, by facilitating the programs and progress of all sectors such as the universities, State government agencies and Federal research organisations, and by acting as an interface between them. This was a novel way of conducting science in Australia, and uncommon even elsewhere in the world. It effectively promoted a 'consortium' approach to marine research, justifiable then and to an even greater extent now, in an era of diminishing resources, as a means of achieving very expensive research in the most efficient possible way.

VIMS immediately embarked upon an initiative which was the exemplar of this novel approach to research. Between 1978 and 1981 it planned and conducted a large, multi-institutional and multi-disciplinary program called the Bass Strait Study, the results of which have profoundly enhanced our knowledge of Bass Strait. The Bass Strait Study was funded by the Victorian Government and drew together, in planning, organising or executing research projects, about fifty scientists from over twenty institutions, including the universities, colleges of advanced education, State agencies, and Federal laboratories. More than thirty projects were initiated, covering physical and chemical oceanography, geology and sedimentology, botany, zoology, fisheries, pollution monitoring and data assessment and management.

The Bass Strait Study wound down in 1981, as funding from the Victorian Government became more tightly constrained. However, events in the Federal sphere during the late 1970s began to exert a significant influence on Australian marine science and helped Bass Strait research maintain the momentum gained through VIMS' Bass Strait Study. Reports to the federal government by its advisory committee, AMSTAC, in 1979, 1980 and 1981 identified Bass Strait as one of the areas which should be given national priority in marine research. Thus, when the Federal government made funding available through the Marine Science and Technology Grants Scheme in 1980, there was a recognition that research in Bass Strait should receive special support. This, coupled with the fact that there was by this time a large group of scientists who (because they had worked in Bass Strait for the previous few years) were poised to prepare proposals and obtain grants from the Federal Government scheme, meant that the progress of knowledge about Bass Strait did not falter.

Thus there have been several factors which have led to greater research activity in Bass Strait: the heightened interest in matters marine, throughout Australia in the 1970s, the formation of research organisations in Victoria, the catalyst provided by the VIMS Bass Strait Study, and the consolidation of much of this effort through the establishment of the MST Grants Scheme. There also have been several other complementary processes. In Tasmania, the effectiveness of the Tasmanian Fisheries Development Authority and its recent successor, the Department of Sea Fisheries, brought additional research efforts to bear on Bass Strait, particularly on the subjects of scallops and squid. The Federal government decisions in the late 1970s to establish the Australian Maritime College at Launceston, and in 1980 to relocate the CSIRO Divisions of Fisheries Research and Oceanography in Hobart, added greatly to the number of researchers and educators on the southern side of Bass Strait. CSIRO responded by developing a number of new programmes in Bass Strait, while the Maritime College operates numerous training cruises in its waters and conducts special courses for the fishing industry.

What also must not be forgotten is the burgeoning interest in teaching marine science, as well as in doing research, that developed in educational institutions in the 1970s. All four Victorian universities now have strengths in marine science in at least some of their departments, while one, the University of Melbourne, had chosen to give a special emphasis to marine science by seeking to co-ordinate efforts across many departments. The University of Tasmania, too, had a long-standing presence in marine science, which expanded during the 1970s. This activity in both States led to a growing pool of expertise, interested in and able to focus upon the scientific unknowns of Bass Strait.

It is clear from the institutional and structural developments outlined above that, for Bass Strait research, the stage was set and the play began. There have been many highlights and some contretemps, and it is difficult to say with certainty, given the vagaries of research funding and the ever-increasing expense of offshore work, whether the climax has been reached or is yet to come. However, in reviewing the achievements of a decade of research, critical accolades are clearly appropriate.

Much of the most intensive (and expensive) research has been in physical oceanography, in determining tides, currents, winds and waves, and understanding the relationship between Bass Strait and adjacent oceanic waters. This effort may be seen at least partly as a response to the needs of the offshore oil industry, outlined elsewhere in this book. As well as some of the organisations already mentioned, the Royal Australian Navy has played a major role, both in its provision of the HMAS *Kimbla* (recently decommissioned) and HMAS *Cook*, and in the efforts of its own research scientists at the Royal Australian Navy Research Laboratories. Through the work of many researchers we are able for the first time to define the tidal patterns in all parts of Bass Strait, and to predict the currents which result from tides and winds. These are now known to be very complex, and could not have been determined without large computer models and extensive measurements by tide gauges and current meters. The results of this work have immense significance for offshore construction, transport, navigation and pollution control.

Some of the large-scale physical processes also are becoming understood. There is a slight nett movement of water through Bass Strait in winter, from west to east, and on the eastern margin between Victoria and Tasmania there occurs the Bass Strait Cascade, 'the greatest waterfall in the world', in which water near the bottom of Bass Strait tumbles about 600 m down the continental slope, because it is slightly denser than the water of the adjacent Tasman Sea. The impressive advances in physical oceanography were summarized at a 'Bass-84 ' meeting in Queenscliff in February 1984, leading to a remarkable, co-ordinated effort during 1984 and 1985 involving scientists from around Australia and overseas.

The supply and distribution of nutrients necessary to support the marine food chains, including commercially valuable species, also is now better known, but this is one area in which many gaps remain in our knowledge. There is still too little systematically collected information, but that which has been gathered indicates that a combination of processes contribute nutrients to Bass Strait water, including upwelling along the western continental slope, intrusions of cold Antarctic water along the north-west coast of Tasmania, and the spilling of eddies of water from the Eastern Australian Current into Bass Strait during the summer. Another process which may prove to be significant is the regeneration of nutrients from bottom sediments caused by the breaking of 'internal waves' about 100 metres below the sea surface. Just like waves on a beach, they may disturb the bottom sediments and release vital nutrients such as nitrogen and phosphorus. However, it does seem that these sources together do not produce an abundance of nutrients, for Bass Strait waters have only low nutrient concentrations, unlike similar extensive continental shelf areas in other temperate parts of the world. At the same time, it is comforting to know that Bass Strait waters are relatively unpolluted, since studies done in conjunction with some of the nutrient measurements indicate that heavy metals, hydrocarbons and pesticides are present only at almost undetectable levels.

The distribution of sediments and other geological features is one of the least imperfect areas of knowledge about Bass Strait. This undoubtedly reflects the interest of the oil exploration companies and of the federal agency, the Bureau of Mineral Resources, Geology and Geophysics, as well as important work by university scientists. Work during the last decade has been concerned with understanding the processes of sediment stability and transport while, more recently, attention has shifted towards the continental slopes, which are deeply dissected by canyons and ravines, exposing outcrops of rocks normally deeply buried under sediment accumulations. An interesting revelation is that the eastern continental slope is slowly eroding from the effects of the Bass Strait Cascade identified by the physical oceanographers.

It is often the animals and plants which attract most interest, and about which most questions are asked. These also often are the most difficult questions to answer, since the immense variety of organisms, some not previously known, are usually very irregularly distributed. However, through intensive efforts at the Museum of Victoria and several Victorian universities, the phytoplankton (microscopic, one-celled plants), the zooplankton (slightly bigger, but still mostly microscopic animals which float in the water and feed on phytoplankton or on one another), the benthic (bottom-dwelling) animals and plants, and the demersal (near-bottom) and pelagic (mid-water) animals, such as fish and squids, are now fairly well known. These studies reveal very complex patterns which result from processes operating on time scales from several days to millions of years.

For instance, the types of phytoplankton vary according to the source of the water mass they occupy, such as cold Antarctic water, upwelling water, or warm, eddying East Australian Current water. The different phytoplankton species may be used to plot the movements and mixing of these water masses over periods of days and weeks. At the other extreme, the communities of benthic organisms which we see today undoubtedly result from changes which occurred during the series of rises and falls in sea level over the last half-million years or so, outlined in the article by Dr Jennings. These communities contain species from the eastern coast of Australia and from Western Australia and the Great Australian Bight as well as endemic species (those restricted to the Bass Strait region). The species composition at any location and the changes between locations present very complicated patterns. It is probable that, during periods of high sea level (inter-glacials), waves of colonization of different species could occur from all directions. However, during the various ice-ages, when sea level was lower, only patches of species survived in deeper water, to be later supplemented by further recolonizations, sometimes by a whole new range of species, as sea-level again rose.

From the preceding summaries, it is possible to conclude that our knowledge of many aspects of Bass Strait has increased dramatically in the time since the first edition of this book. There even are efforts to make much of this exciting information available to the general community, through development of educational facilities such as the Marine Studies Centre operated by VIMS at Queenscliff. Activity continues apace, largely supported by new funds from the MST Grants Scheme to university and museum researchers, as well as by bodies such as the CSIRO Divisions of Fisheries Research and Oceanography, the Marine Resources Management Branch in Victoria and the Department of Sea Fisheries in Tasmania. Much of the most recent work concerns living communities, particularly the commercially important species they contain, in deeper waters off the continental slope. As well, new, detailed hydrographic surveys of the bathymetry of Bass Strait and adjacent areas in the last couple of years are now becoming available to support researchers, and many new data have been collected on tides and currents to complete the hydrodynamics picture.

However, there remain many gaps, some big, some small and too numerous to mention, in our understanding of Bass Strait. The large ones have been alluded to, such as the uncertainty about what drives the nutrient cycles in Bass Strait, and how this affects natural production and, ultimately, fisheries yield. Although the large scale movements of water in Bass Strait are better understood, small-scale patterns so important to environmental management and pollution control are not yet clear. The problem of planning and regulating the use of coastal resources, including the declaration of marine reserves and parks, is only just beginning to be faced in the States bordering Bass Strait, and research of various types will be required to support such developments.

Victorians and Tasmanians, and indeed all Australians, therefore cannot be complacent about the need to do more, and to know more, about Bass Strait. Nonetheless, we should be prepared to make the transition from confronting the last frontier to wisely managing a better-known and valuable terrain.

# 9
# WILDLIFE OF BASS STRAIT

**Jeannette Hope**

When European explorers first sailed into Bass Strait and saw the rugged and mountainous islands, their immediate impressions were of sterility.

'The whole of this land is exceedingly barren . . . It is a barren spot . . . It is a most uninteresting spot.' So they described the islands recognised today as possessing some of the most spectacular scenery and fascinating wildlife in Australia.

When they were first explored, at the end of the eighteenth century, the islands were uninhabited by Aborigines; none of the smokes of Aboriginal fires, so common along the Tasmanian coast, were seen on the islands. To Matthew Flinders, the islands seemed 'shunned by almost every kind of animated being'.

This attitude was probably due to the strangeness and unfamiliarity of the landscape to the early explorers for, when they did see animals in Bass Strait, these men were astonished by their vast numbers. Here is Matthew Flinders' description of a flight of mutton-birds, or as he called them, sooty petrels, as seen on December 9, 1798:

> There was a stream from 50 to 80 yards in depth, and of three hundred yards, or more, in breadth; the birds were not scattered but flying as compactly as free movement of their wings seemed to allow; and during a full hour and a half, this stream of petrels continued to pass without interruption, at a rate little inferior to the swiftness of a pigeon. On the lowest computation, I think the number could not have been less than a hundred millions.

Petrels and penguins, gannets and gulls, albatrosses and cormorants, all these nested in large rookeries on the islands. Cape Barren geese, swans and many other birds were also seen, and on the rocky headlands and islets were hundreds of seals.

Apart from the teeming sea bird rookeries and the abundant seals, the islands were also inhabited by many of the familiar animals of Tasmania and of the mainland — wallabies, wombats, bandicoots, tiger cats and spiny anteaters, as well as the emu and many other land birds, some peculiar to Tasmania and the islands, and a variety of snakes and lizards.

Terrestrial animals like these usually reach islands,

The famous gannet rookery on Cat Island in the Furneaux Group in 1909. Depredations by fishermen, using gannets for bait, have reduced the colony on Cat Island to no more than a few breeding pairs today. Attempts are being made to re-establish the rookery.

*Cape Barren Geese—the archetypal Bass Strait symbol.*

especially oceanic ones hundreds of miles from the nearest continent, by flying or by drifting across the ocean on floating rafts. But in Bass Strait land animals did not reach the islands by these methods. They were already living there, on a land bridge connecting Tasmania to the Australian mainland, and were stranded on the islands by the rising sea level at the end of the last ice age.

The last two million years of earth's history, the Pleistocene period of geological time, was marked by extreme climatic changes as the ice ages waxed and waned. World sea level rose and fell as the polar ice caps melted or expanded. During the colder times, glaciers formed on the mountains of western Tasmania and on the highlands of southern New South Wales. As sea levels fell, a corridor of dry land linked Tasmania, through the eastern island chain, to Wilson's Promontory in Victoria and, at the lowest sea level, about 120 m below present, a western corridor through King Island was also available. The modern islands were mountain and hills far inland. When the sea level rose again and covered the land bridge, some animals and plants were stranded on hill tops — now the islands of Bass Strait. The last rise in sea level began about 18 000 years ago; by 12 000 years ago, Tasmania was severed from the mainland, and by 5000 years, the islands had reached their present size.

What was the country like when the sea was lower and it was possible to walk to Tasmania? At one time, probably about 30 000 years ago, cool temperate rainforest grew at least on King Island. This included the Tasmanian myrtle or beech (*Nothofagus cunninghamii*), which is not now found on any island. From the peaty swamps of King Island have come fossils of *Diprotodon* and *Zygomaturus*, giant marsupials the size of a rhinoceros. Bones of extinct kangaroos, and of a large (up to one metre long) anteater or echidna, similar to those now found only in New Guinea, are also known from King Island. These extinct animals were widespread throughout Australia in the past, but had died out by about 20 000 years ago.

Fossils have also been found in island caves. Ranga Cave, on Flinders Island, contains the bones of the grey kangaroo, called boomer or forester in Tasmania. This species lived on Flinders Island until at least 8000 years ago, when the island had just separated from Tasmania. Also living on Flinders Island at that time were the eastern native cat (*Dasyurus viverrinus*), the barred bandicoot (*Perameles gunnii*) and a rat-kangaroo (*Aepyrymnus rufescens*), now found only in northern New South Wales and Queensland. None of these species are now found

88

on the islands, but the grey kangaroo, native cat and bandicoot still live in Tasmania.

Perhaps the best record of the past comes from Cave Bay Cave, on Hunter Island. In this cave, layers of sediment which built up between 23 000 and 15 000 years ago, when the sea level was lowest and the climate coldest, contain a remarkable sequence of plant pollen and animal fossils. The pollen record shows that much of the western Bassian peninsula at that time was covered with grassland and daisy fields. The closest modern parallel is the high treeless plains of central Tasmania. The commonest mammal in Cave Bay Cave was the broad-toothed rat (*Mastacomys fuscus*) which now lives in small numbers in patches of buttongrass in Tasmania. Most of the other mammals present during this time were animals of open country. A surprising find was the toolache wallaby (*Macropus greyi*) a species now extinct, but historically occurring in open grassland in western Victoria and southeast South Australia.

Down the western and eastern sides of the land bridge were corridors of open country, along which grassland species such as the toolache wallaby and rufous rat-kangaroo could move. The lower country in the centre of the land bridge was probably a large lake, or at times an estuarine embayment, while the scrub and forest-covered hilltops may already have been ecological 'islands' long before the rising sea level turned them into real ones.

After the sea had reached its present level, many of the new islands could not support the range of species originally present, because of the limited areas available, and a gradual simplification of the fauna took place. By 1800, while the larger islands still supported a wide variety of plants and animals, the small ones were inhabited by very few species. For example, though the tiger snake (*Notechis ater*) and the copperhead (*Denisonia superba*) both occur on Flinders Island, on the smaller islands nearby only one or the other occur, never both together. Similarly, the kangaroo and wallaby are not found together on any island of less than 400 ha.

The kangaroo and wallaby, as they are called in Tasmania and on the islands, are respectively the red-necked wallaby, *Macropus rufogriseus*, and the Tasmanian pademelon, *Thylogale billardierii*. Before 1800 these two species inhabited many islands in Bass Strait. The larger red-necked wallaby was restricted to the larger islands, such as Flinders, King and Cape Barren, and still survives on most of them. But the little pademelon originally lived on almost every island more than 120 ha in size, about twenty islands altogether. Continual hunting has caused it to die out on many, and today it is found on only eight or nine islands. This pademelon died out on the Australian mainland before 1900, and is now found only on the islands and in Tasmania.

Early Aboriginal people may have been affected in a similar way as the sea levels rose. They had walked to Tasmania when the sea level was low, and had penetrated into the wilds of the Franklin River by 20 000 years ago, when that area was experiencing icy alpine conditions. Aborigines lived on Hunter Island at that time too, and hunted the toolache wallabies and other large animals whose bones are found there (but not the rats; these were brought to the cave by owls).

Aboriginal groups remained on at least King and Flinders Island after the sea level isolated these, because archaeological sites dating to about 7000 years have been found. Yet most of the islands were uninhabited and unvisited by Aborigines when Europeans first arrived. What happened to the people there? Did they leave the islands, or did they die out? Perhaps even King and Flinders were just too small to support a viable human population over thousands of years.

When Europeans arrived in Bass Strait, the animals on the smaller islands succumbed easily to hunting pressure. One characteristic of island populations is their lack of timidity. It was possible for the sealers to walk right up to nesting gannets and albatrosses and kill them, and for the Aboriginal women who lived with the sealers to crawl among the seals before clubbing them to death. The sealers brought dogs with them to hunt kangaroos, and these had a drastic effect on the mammals on the islands where few predators had previously lived. On some islands the animals are still amazingly tame, and it is possible to walk among nesting sea birds. On Deal Island, in the late afternoon, dozens of red-necked wallabies graze on open grassland around the light-station buildings. It is possible to walk up to within a metre or so of the wallabies before they move slowly away.

Many animals and plants were collected on the islands by early explorers long before they were known to live on the mainland. These include the wombat, and one of the first specimens of this animal was collected on Cape Barren Island by George Bass in 1798, when he and Matthew Flinders circumnavigated Tasmania in the *Norfolk*. At that time wombats lived on several islands; King, Flinders, Clarke, Cape Barren and possibly Deal and Badger. On King Island in 1802, several naturalists from the French expedition led by Baudin were marooned for a few days. Here they were hospitably entertained by some English sealers, the leader of whom was a man named Cowper. Cowper showed the Frenchmen two emus hanging in his larder, and related how he had domesticated wombats, training them to go out and feed during the day and to return to their huts at night. Unfortunately this is the only account we have about the emu and wombat on King Island. Both were varieties unique to that island, and both are now extinct there. By the 1880s, when the Field Naturalists' Club of Victoria sent several expeditions to the islands, wombats appeared to have been wiped out on all the islands. Fortunately they survived on Flinders Island and are fairly numerous there, although they are regarded as a pest by the farmers.

Some species have been unaffected by the last 170 years of human settlement on the islands, often because of the isolation and inaccessibility of certain islands. The albatross rookeries on Albatross Island and the ancient forests on Rodondo Island are close to their original state, because both these islands are bound by sheer cliffs which make landing very difficult. But on the more accessible islands, wildlife survives only because of sheer numerical strength, as, for example, in the case of the mutton-birds, or because some islands, such as Flinders, were not settled until fairly late and also have large areas of mountainous country which act as refuges.

The effect of man is twofold: directly he acts on the wildlife by shooting and poisoning, and on the vegetation by burning and clearing; but indirectly, fires and land development have an unpredictable long-term effect. One of the species that has suffered badly by direct hunting is the gannet of Cat Island, one of the only five gannet islands in Australia. Fisherman found these birds a convenient source of bait because they are so easy to catch. In 1893 there were 2500 gannets nesting on Cat Island. By 1935 only 500 pairs were left, and the Tasmanian Government, through the Animals and Birds Protection Board, put a warden on the island to protect the gannets. When shortage of money meant that no warden could be employed, the gannets decreased in numbers until now few remain. Cat Island is now a wildlife sanctuary.

Early in the nineteenth century men flocked to the islands to kill the seals, which were valued for their fur and oil. Between 1800 and 1806 about 100 000 seal skins from Bass Strait passed through Sydney, and by 1820 the seals were almost completely exterminated. Three different seals were originally present in

At the 'Pole of Inaccessibility' of Bass Strait, Southwest Island lies some ten miles off the Kent Group, close by Matthew Flinders's Judgement Rocks, one of the great seal colonies of Bass Strait. Ancient rock shelters are found near the peak of this island, perhaps the homes of sealers in the early days.

Bass Strait; the elephant seal (*Mirounga leonina*), the hair seal (*Neophoca cinerea*) and the Australian fur seal (*Artocephalus pusillus*); but only the fur seals still live there; though relatively few, they are still considered a pest by fishermen.

As the seals diminished in numbers, the sealers found that the large numbers of kangaroo and wallaby on the Bass Strait islands made a profitable substitute. In 1830, one man reported that on Hunter Island; 'the 11 months he was there he saw 4000 wallaby skins and thought nothing of going out of a moonlight night and catching 40 or 50'.

The great Australian sport of lighting bushfires has had a long and inglorious tradition on the Bass Strait Islands. John Lort Stokes, who explored Bass Strait in the *Beagle* in 1838-39, casually noted in his journal: 'The fire that had been accidentally kindled on Three Hummock Island when we were last there (three weeks earlier), was still burning.'

The vegetation on many islands had been completely altered by burning. The south-eastern part of King Island was originally covered with a magnificent forest of blue gum, up to ninety metres high. Now the whole area is covered with grassland, and all that remains of the forest is a few blackened tree trunks and one or two living gums in isolated gullies. It is clear that a great variety of plants once grew on the islands. On some islands, of 400 ha or less in area, hundreds of species have been found, while on Flinders Island, for example, something like eighty different orchids have been recorded, although many have not been seen recently. Unfortunately it is something of a race with the land developers for the botanists to discover what is still there, let alone to try and conserve some of it. It is even more difficult to determine what grew on the islands before 1797, when settlement began. For example, the Tasmanian beech, we know, grew on King Island 30 000 years ago, and may well have lived there before the island was extensively burnt during the last hundred years. Perhaps it still grows in some sheltered valley in the south-east of King Island.

Until the last forty years or so, the islands of Bass Strait had a very small human population and, with a few exceptions, such as the gannets, those animals and plants that had survived the first impact of man in the early nineteenth century had generally adjusted to his presence. But with the Soldier Settlement Schemes on King and Flinders Islands the population has grown and extensive land development has taken place. This will certainly have a long-term effect on the wildlife, and many of the smaller and rare animals, such as the potoroo and tiger cat, will eventually die out.

There is some hope, however, in the recent establishment of many conservation areas in the islands. The Strzelecki Ranges on Flinders Island are now a national park, and other tracts of land on Flinders and King have been set aside as state reserves (Kentford Forest on King, an area of relict eucalyptus forest), nature reserves or wildlife sanctuaries (Lavinia, Reekara and Sea Elephant River on King Island; Logan Lagoon, Lackrana, Badger Corner and the Patriachs on Flinders Island.) In addition, many small islands and islets are nature reserves or bird sanctuaries, specifically set aside for seabirds and seal breeding, or for the Cape Barren Goose. Others are game or mutton bird reserves. Recently Trefoil Island in the Hunter Group, one of the major commercial mutton birding islands, was purchased by the Federal Government for use as a co-operative by the Aboriginal descendents of the sealers and Aboriginal women who began the industry nearly two hundred years ago.

For a country its size, Australia has very few offshore islands. Of these the islands of Bass Strait are among the most important, both historically and biologically. Islands, too, are particularly suited as reserves; they are clearly defined in area and can be easily guarded, they possess unusual wildlife, and they have a great fascination for most people. It is regrettable that no island of any size has been preserved in its entirety as a National Park or Nature Reserve. Cape Barren Island, for example, would be one possibility, especially since it has been nominated as a wilderness area to the Australian Heritage Commission.

**Further reading:**

HPC Ashworth and D Le Souef, 1895: 'Albatross I and the Hunter Group', *Victorian Naturalist*, 11:134-144

AJ Campbell, 1888: 'Field Naturalists' Club of Victoria Expedition to King Island, Nov. 1887' *Victorian Naturalist* 4:129-164

J Gabriel, 1894: 'Report of Expedition to Furneaux Group', *Victorian Naturalist* 10:167-184

Hope JH, 1973: 'Mammals of the Bass Strait Islands' *Proceedings of the Royal Society of Victoria*, 85:163-196

Hope, JH, 1984: 'The Australian Quaternary', pp 69-81 in *Vertebrate zoology and evolution in Australasia*, ed M Archer and G Clayton, Hesperian Press, Perth

Jones R (ed.), 1981: 'Exploited and endangered wildlife: population management of wallabies, possums, muttonbirds, Cape Barren geese and forester kangaroos in Tasmania', *Centre for Environmental Studies Occasional Paper* 12, University of Tasmania

JH Mullett and S Murray-Smith, 1967: 'First footing on a Bass Strait Island: an investigation of Dover Island in the Kent Group' *Victorian Naturalist* 84:239-250

Russell JA, JH Matthews and R Jones, 1979: 'Wilderness in Tasmania: a report to the Australian Heritage Commission' *Centre for Environmental Studies Occasional Paper*, 10, University of Tasmania

# 10
# PETROLEUM IN BASS STRAIT

## Rick Wilkinson

Like the search for gold there is always a latent excitement about oil exploration, an excitement which will suddenly burst into the open when discoveries are made, particularly in a new province. Recently that excitement has been seen in the deserts of South Australia and Queensland and in the isolated waters of the Timor Sea, but back in the mid-1960s the country's eyes were rivetted to a narrow stretch of water between the south-east coast of Victoria and Flinders Island.

In February 1965 the 5500 tonne drillship *Glomar 3* struck gas in a well called East Gippsland Shelf 1. It was the very first well to be drilled in the region and it caught the oil explorers (Esso/BHP) and the whole country completely by surprise. An appraisal well confirmed the discovery four months later and the field was called Barracouta, following the explorers' decision to name this and their future wells after species of fish. In fact Barracouta was destined to be the beginning of such an amazing series of successes that by 1970 Australia was seventy per cent self sufficient in oil, Victoria had natural gas reserves to last into the twenty-first century, and Esso/BHP were catapulted into the forefront of offshore engineering technology. In just five years Bass Strait had earned a place on the world oil and gas map.

Yet, despite the apparent ease of this 'dream run' of discoveries, the build-up to the exploration progamme was a gradual one dotted with missed chances, conjecture, coincidence and more than a share of luck. As early as the nineteenth century there were reports of bitumen deposits being washed up on beaches at the western end of Bass Strait and their origin was the subject of debate right up till the 1960s. The old South Australian school of thought suggested either a South American or Antarctic origin, while the oncoming younger geologists were more inclined to believe the deposits originated from local sources. Although the debate waxed and waned over the years it did focus attention on Bass Strait from the beginning, and prompted early drilling in the Otway Basin of South Australia and western Victoria.

In the Gippsland Basin of Eastern Victoria oil seeps had been known onshore for some time. These prompted the formation of several oil companies in the 1950s, the most lasting of which is Woodside Petroleum (known originally as Woodside (Lakes Entrance) Oil. Woodside began with an exploration permit near the Ninety Mile Beach and took its name from a small farming township in the district. Drilling there was never commercially successful, but more than once the company geologists and visiting experts pointed at the waters of Bass Strait, suggesting that viable oil reservoirs could be found offshore.

At the time Woodside's finances were perilously low, to say the least, and on top of that the expertise involved in offshore exploration was in its infancy, even in the USA. After further urging the company did apply for and receive a small offshore permit close to the Gippsland coast in 1959, but could not afford any greater risk in deeper waters. A gas find called Golden Beach in this permit area ten years later shows how close the company had been to a major prize.

Also at work on the fringes of Bass Strait in the 1950s was a company called Frome-Broken Hill, an amalgam of three oil majors (BP, Mobil and Esso) and the Australian Zinc Corporation. Although they were in partnership the rivalry between the majors was still severe, and in fact the company broke up in the early 1960s. However the point is that all three oil companies had a knowledge of the data gleaned from the work around the Strait and, of the three, Esso was the keenest to try an independent look at the offshore region—particularly the Gippsland Basin.

It was at this point that the legendary Lewis Weeks arrived on the scene. A US geologist with a sound understanding of global geology, Weeks had recently retired as chief geologist for Esso, and so had access to the Bass Strait data from that company's part in the Frome-Broken Hill group. The story of his being invited to give an opinion on the petroleum potential of the Sydney Basin by the Australian iron and steel company BHP, his negative comments of that region and subsequent recommendation of the offshore Gippsland Basin in Bass Strait, is now well known.

Weeks knew of the onshore Gippsland oil seeps and he likened these to the Tertiary age rocks in South East Asia, which had already proved to contain viable oil reservoirs. He believed the sandstones which contained minor oil onshore Gippsland would be thicker and better reservoirs offshore. He also knew that offshore technology—particularly in the Gulf of Mexico—was progressing swiftly, and that drilling was possible in Bass Strait.

BHP at that time knew virtually nothing about oil exploration, but the company acted quickly on Weeks' advice, taking out permits from the Victorian, South Australian and Tasmanian Governments. These covered some 163 000 sq km in the offshore Gippsland, Otway and Bass Basins, and took up most of the available acreage in the Strait. With Weeks' continued guidance BHP contracted the Canadian Aero Service company to fly an aerial survey to give a basic outline of the basins and their sediments by measuring local changes in the earth's magnetic field, using an instrument called a magnetometer. The survey results were encouraging and Weeks further recommended a marine seismic survey over the Otway and Gippsland Basin leases to fill in some of the detailed geology. It was the first survey of its type in Australia and cost well over $1 million (£650 000 in currency of the day) using two vessels brought in from Italy. The technique involved setting off explosions in the water along pre-arranged grid lines and recording the frequency of the shock waves after they had bounced off the rock formations beneath the seabed. The results indicated a number of potential oil structures, particularly in the Gippsland Basin, and BHP decided to make that area its top priority.

At that point the Australian company began to look around for a partner with both the cash reserves and the technical oil expertise to continue the exploration program in detail. The farm-in package was worked out in BHP's Melbourne headquarters in 1964, and the terms were tough even by today's standards.

The incoming partner would have to pay the cost of all exploration and take the risks for an indefinite period. BHP would retain a straight fifty per cent of anything found, and its future costs would not begin until the production program stage. Only at that point would BHP pay half the production costs and recompense half the exploration costs leading up to that particular find. (It added the proviso that if it did not want to become part of the production team it would take 12.5 per cent royalty instead, but in fact this proviso has never been acted upon.)

More than twenty companies—reading like a who's who of the world oil industry—were interested in a joint venture with BHP, although the list was shortened to ten invitations to tender. Most rejected the terms as too strict. The main exception was Esso, which had been actively pursuing its own studies in the Gippsland area since the Frome-Broken Hill days. In the event Esso's bid was accepted, and the formal agreement between the two companies was signed on May 12, 1964 for exploration of

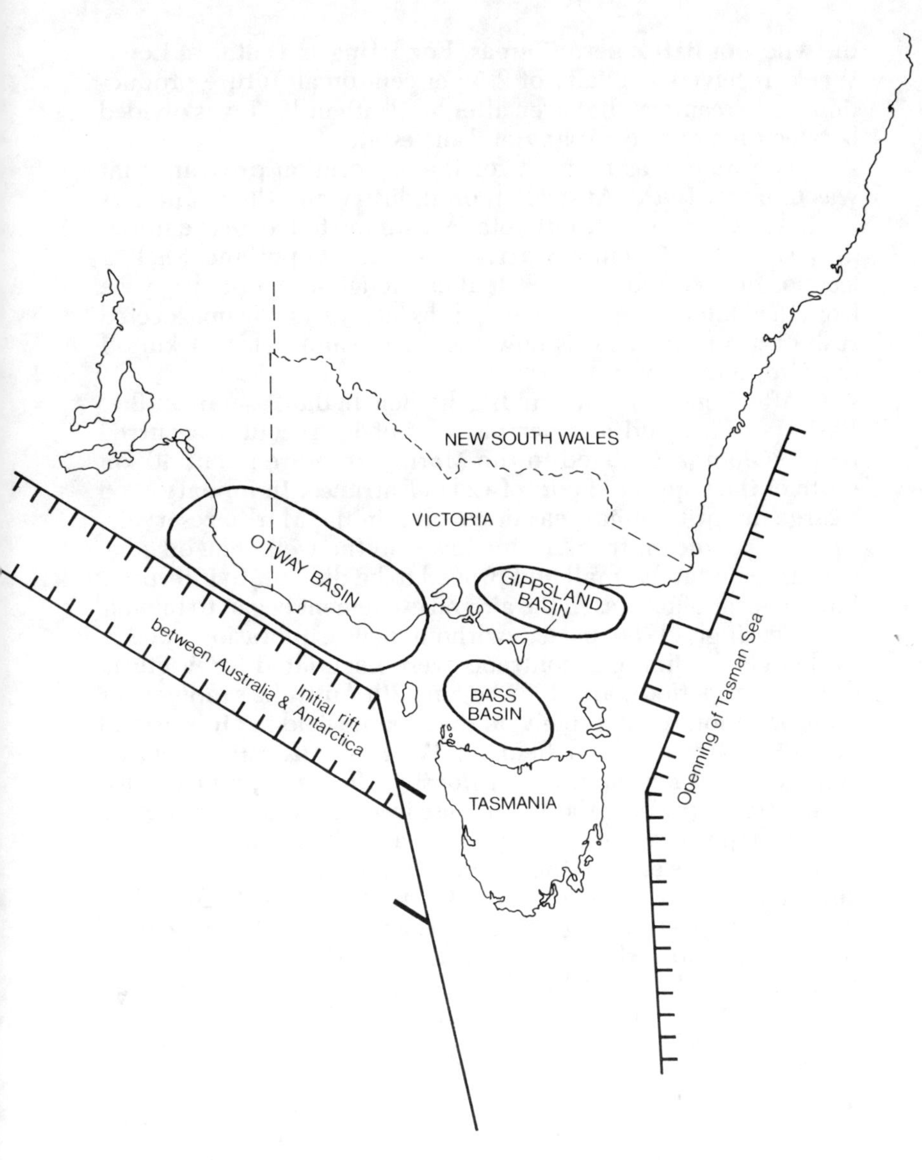

NEW SOUTH WALES
VICTORIA
OTWAY BASIN
GIPPSLAND BASIN
BASS BASIN
TASMANIA
between Australia & Antarctica
Initial rift
Openning of Tasman Sea

the whole of BHP's permit areas. For acting as a catalyst Lewis Weeks received a royalty of 2.5 per cent on all future production — a percentage that even after his death in 1977 was divided between his own company and his estate.

The stage was now set for the exploration program that was to electrify the Australian oil industry and the country as a whole. After a short, but detailed, survey to locate the initial drill sites, the *Glomar 3* arrived at East Gippsland Shelf 1 location on December 27, 1964, after a rough seventy-day voyage from the Gulf of Mexico. Two months later came the unexpected rush of gas from what is now the Barracouta field, 24 km off the Gippsland coast.

After confirming the find, a dry hole in the Tasmanian Bass Basin permits, and another just east of Barracouta, occurred before *Glomar 3* moved to the Marlin structure about 40 km south of the Gippsland port of Lakes Entrance. In January 1966 a large quantity of gas was discovered in the Marlin reservoir, as well as some oil from a zone lower in the well. The oil zone was unfortunately small and proved to be difficult to interpret, but the scent of oil was in the air and excitement began to mount.

Teasingly 1966 passed without new exploration success (although the first gas contracts were negotiated for Victoria during that period), and 1967 began with Australia's indigenous oil production relying solely on the Moonie field in Queensland and the Barrow Island field off Western Australia declared commercial a few months previously. *Glomar 3*, still in Bass Strait after appraising the Marlin discovery, moved to a structure called Kingfish in April and spudded a well on what Esso/BHP thought was the crest of the feature. It discovered a thin oil zone and at face value was only moderately successful. But after studying the well data geologists began to suspect the location was on the limb of the structure and not on the crest at all. They decided to drill a second well, but in the meantime the rig had moved on to a nearby feature named Halibut.

At the end of May 1967 geologists on board the vessel could hardly believe their eyes when oil flowed up the Halibut 1 well at a huge rate during the test program. The drill bit had tapped one of the most perfect reservoirs anywhere in the world. The reservoir rock was very porous and the permeability (a measure of the passages in the rock connecting the individual pore spaces) excellent. There was virtually no gas, and such a strong water drive from beneath that oil was pushed out of the ground at a high rate without mechanical help. Later production work showed the field was capable of producing at over 200 000 barrels a day, but even without that appraisal data geologists sitting on the first well had no hesitation in declaring Halibut a commercial field on the strength of information from that well — a very rare practice in the oil industry.

With spirits soaring *Glomar 3* was sent back to Kingfish and the partners' joy knew no bounds when the number 2 well — this time accurately pin-pointing the structural crest-tapped a huge oil reservoir. Kingfish 3, drilled soon after, confirmed the find, which has subsequently proved to be Australia's biggest field to date and ranked in the top twenty offshore fields in the world.

Australia had finally found enough oil to eliminate its dependence on overseas supplies for many years to come, and by the end of 1968 Esso/BHP was confidently predicting a flow of 135 000 barrels a day by 1970. Undoubtedly the 1967/68 period was the high point of Bass Strait exploration and, although further discoveries were made and are still being made, the later work has come as something of an anticlimax after such an unprecedented run of early discoveries. The initial Gippsland success continued into the early 1970s with smaller oil finds at Tuna and Mackerel, gas at Snapper and marginal oil at Cobia, Flounder, Bream, Turrum, Perch and Dolphin. Plans were laid for the development of Mackerel, Tuna and Snapper, and by that time a production rate of 350 000 barrels a day was considered a probable average rate for the decade. At the same time Victoria was assured of natural gas well past the year 2000.

However commercial success was completely lacking in the Bass and Otway Basins and, despite teasing non-commercial finds at structures like Bass, Pelican and Cormorant, Esso/BHP gradually withdrew from these permits during the 1970s. Other companies have revived activity in both regions from time to time, but as yet they have found nothing commercial either.

Exploration work in the Gippsland Basin also suffered a lapse during the mid-1970s, but this was more to do with what the oil industry saw as unfavourable Federal Government oil policies than with the lack of prospectivity. By 1977 this phase had passed and 1978 saw the first of a new wave of discoveries off the Gippsland coast that has continued intermittently until the present time.

The first came in August 1978 when the Australian built semi-submersible rig *Ocean Endeavour* discovered oil in Seahorse 1, about mid-way between the Snapper and Marlin gas fields. Although a small marginal structure, Seahorse was hailed as an important find because it was the first under the Federal

Government's newly announced 'new oil' policy which priced the country's finds after 1976 at world prices (adjusted for freight and quality differentials) and free of any levies.

However the following month saw a much more important discovery — one that in 1986 is still the biggest of the second wave of Gippsland Bass finds. Also drilled by *Ocean Endeavour* it began life as the West Halibut structure. The reason was that geologists and geophysicists saw the feature as an extension of Halibut which could not be reached from the existing Halibut production platform. The well West Halibut 1 not only penetrated a good oil zone, but also raised the possibility that it might in fact be geologically separated from Halibut. The importance of this to Esso/BHP was that if it was indeed separate from Halibut then it would qualify for the 'new oil' price.

The structure's name was quickly changed to Fortescue and three appraisal wells were drilled in quick succession. The net result was an increase in Bass Strait reserves of around 300 million barrels of recoverable oil and geological proof of the independence of Fortescue from Halibut. This proof was submitted to and accepted by the Bureau of Mineral Resources in Canberra, and Fortescue was officially proclaimed 'new oil' in October 1979.

News of the BMR's acceptance of Esso/BHP's proof of Fortescue as a separate structure from Halibut had three immediate effects. First it prompted the partnership to go ahead with a planned multi-million dollar development plan which involved the development of Fortescue and three previously marginal discoveries — Cobia, West Kingfish and Flounder. Second it gave Esso/BHP confidence to announce a new thirty-well exploration drilling program in its Gippsland Basin leases for the four years to 1985. The third thing was that it created a spur for the Victorian Government to announce a new licensing round of exploration permits in the Gippsland Basin.

During the normal course of its programs in Bass Strait Esso/BHP was required to relinquish fifty per cent of its exploration acreage every four years. The partnership had complied with this Government regulation throughout the 1970s, retaining only a core of permits around its proven production acreage. With the upsurge of renewed exploration vigour at the beginning of the 1980s, following the Federal Government's new pricing policy and the subsequent admission of the Fortescue 'new oil' proof, there was a lot of interest from other oil companies in the relinquished Esso/BHP areas. The Victorian government decided to re-gazette most of this acreage in a parcel of four permits. Such was the renewed fever of the time that most of the world's major oil companies and a host of smaller Australian explorers submitted applications. The enthusiasm was fuelled ever further when it was learnt that Esso/BHP itself had applied for one of the new permits, suggesting that there were some attractive structures yet to be explored.

The successful groups in this new licensing round were headed by Shell, Phillips Petroleum and Aquitaine (the fourth permit did not receive sufficient bids to warrant renewal), and they immediately began exploration programs. They were joined by a group headed by the Australian subsidiary of Canadian Hudson's Bay Oil and Gas Company, and another led by the US company Union Texas, both of which had gained relinquished Esso/BHP areas in the late 1970s.

Put with Esso/BHP and its promised thirty wells it meant that there were six groups drilling, or about to drill, in the Gippsland Basin during the first two years of the 1980s. It seemed at the time that Bass Strait still retained some of the exploration magic of the 1960s. There was even a spin-off of enthusiasm for the Otway and Bass Basins, with new groups trying their luck in those areas. However the enthusiasm was short-lived, and it says much for the thoroughness of Esso/BHP that the partnership is the only group to emerge from the most recent exploration round with any prospects for new commercial development.

Some of the other groups — notably Shell (at Basker and Manta) and Hudbay (at West Seahorse, Baleen and Sperm Whale) — did make discoveries, but at the beginning of 1986 they remain on the marginal lists. The other consortia have relinquished their acreage totally although, in a reversal of this trend, Esso/BHP farmed back into the Shell group's permit in 1985 and made the interesting Kipper gas condensate discovery in April the following year. Unfortunately it, too, is non-commercial at this stage. Nevertheless optimism remains and, despite the plunge in world oil prices during 1986 which has severely cut exploration programs everywhere, the Victorian Government began a new licensing round in July 1986.

The other point to be made is that even Esso/BHP could not match the discoveries of the 1960s, and it has taken many months of negotiating with the Federal government over revised oil pricing to set up this most recent development program. Old marginal finds like Bream, Dolphin, Perch and Turrum, along with more recent marginal prospects like Seahorse and Tawhine, and easterly extentions of fields already on stream like Tuna,

Kingfish and Barracouta, form part of the development package announced by the partnership in October 1984. Of these East Tuna, with a possible fifty million barrels of recoverable oil, is the largest prospect while Bream, a possible thirty-five million barrels, is in second place. It is a far cry from the billion barrel plus reserves at Kingfish and 800 million barrels at Halibut estimated after their appraisal, but it does illustrate the way development programs in the Strait have altered since the heady days of 1969.

Much of the activity during the last years of the 1960s was related to the huge job of constructing offshore production platforms, laying pipelines to shore, as well as onshore lines to the gas plant/oil stabilising units at Longford near Sale in Gippsland, and an onshore oil line to Long Island Point terminal further along the coast, plus a separate line under Port Phillip Bay to refineries west of Melbourne. There were also the tasks of setting up a supply base and construction yard at Barry Beach, also on the Gippsland coast.

The offshore construction work was at the frontiers of technology at the time, with the only parallel in water depth and storm potential being development of the southern North Sea gas fields off the United Kingdom and the Netherlands about the same period. The oilmen soon discovered what fisherman had known for generations. Bass Strait can change without warning from a dead calm to a white-capped frenzy within the space of half-an-hour, and it says a lot for the developing engineering technology that the first fields came on stream in 1969, while the initial development of four fields (Barracouta, Marlin, Halibut and Kingfish) was successfully completed by the end of 1971 at a cost of $300 million.

There were two scares during those early years — one a fire and the other a gas blowout on the Marlin platform — but in each case the danger was rectified without loss of human life or any great setback in production schedules, despite the drama they evoked.

The next development phase was spread over the 1970s with a production platform on Mackerel in 1977, one on Tuna in 1979 and another on Snapper completed in 1981. It was noticeable with these three platforms that Esso/BHP had changed to modular design methods which meant that most of the topside facilities were constructed onshore and then placed on the steel platform legs offshore in a series of 'big lifts' with a crane barge. Growing experience with the fickle Bass Strait weather caused this change, as engineers soon learnt that delays could be

# ESSO/BHP's
# BASS STRAIT PRODUCTION SYSTEM DEVELOPMENTS

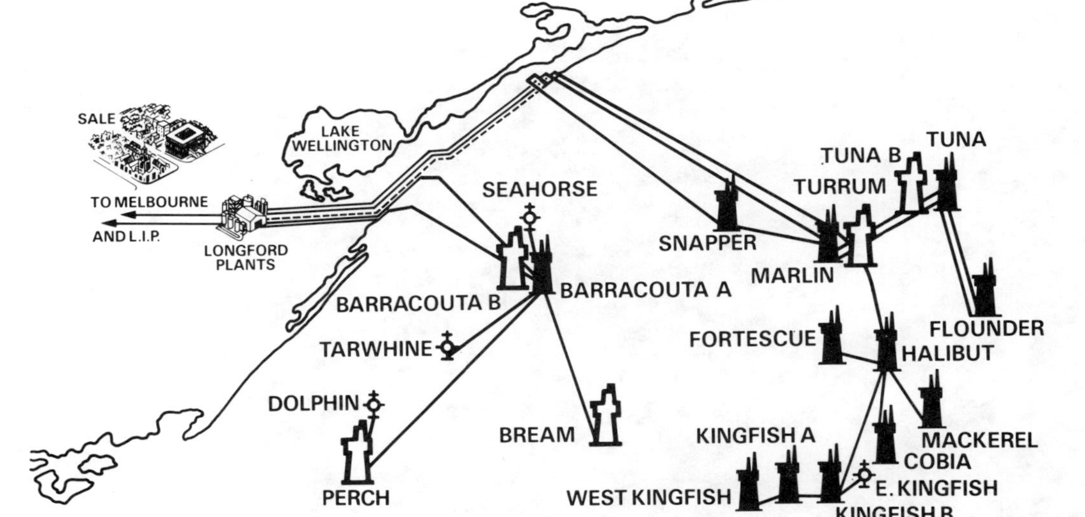


*Esso/BHP's Bass Strait Production System Developments*
*The map illustrates the developments Esso and BHP hope to undertake in Bass Strait over the next five years. The bold platform outlines are the oil and gas production platforms now in place, while the platform outlines are the structures the two companies hope to build. The sub-sea completions will be well-head production facilities installed on the sea bed and linked to the nearby platforms by pipeline.*
*October 23, 1984*

Aerial view of Barry Beach construction yard, Gippsland, Eastern Victoria, March 1981. West Kingfish jacket nearly completed with Cobia jacket beginning construction behind it. Courtesy Val Foreman Photography.

Esso/BHP's Cobia platform ready for launching in Bass Strait. (1982)
Courtesy Val Foreman Photography.

*Barracouta Platform.*

minimised by cutting back on the number of hours a barge needed to be moored beside a platform for the construction process.

The third development period began in the early 1980s when Esso/BHP was given the all-clear on Fortescue oil price and platforms were ordered for West Kingfish, Cobia, Fortescue and Flounder, at a total development cost of close to $1 billion. Normal inflation and the rising sophistication of offshore technology accounted for the big cost increases during the 1970s. Two new elements in technology were the placement of a sub-sea completion unit at Cobia (production via a wellhead in the sea bed rather than on a platform), and the use of high angle directional drilling to reach outlying parts of the same reservoir or an adjoining reservoir. Both these techniques will be employed to an even greater extent in the most recent development programme announced in October 1984.

The return in terms of reserves will be much lower than with the early developments, but Esso/BHP's pricing arrangements with the Federal government should ensure that an adequate financial reward will accrue from the estimated $1.8 billion still hoped to be spent on the new projects before 1990. The companies expect 150 million barrels to be won from these previously marginal finds; at 1984 prices this would have been worth around $5 billion.

It can be seen that government influence, particularly in the form of pricing policies, has played a significant part in Bass Strait oil development. But equally the Federal government's vested interest in the volume of production has grown markedly over the last fifteen years with the rapid rise in world oil prices, and amounted to as much as $4 billion a year from royalties and levies during the early 1980s.

The original Bass Strait leases were issued during Sir Robert Menzies' prime ministership, but it was not until John Gorton's time that the question of oil pricing arose in the region. Negotiations between Esso/BHP and the Prime Minister began in 1968 soon after the discovery of Halibut and Kingfish, and eventually resulted in two pricing arrangements. The first was an interim figure for Gippsland crude of $A2.42 a barrel, to run from the date of first production until September 1970. The second was to run for the following five years and was fixed at the October 1968 world posted price, with adjustments for freight and relevant discounts. That worked out at $A2.06 a barrel, and Esso/BHP were well pleased; they had reckoned on the world price falling in the years ahead, and consequently had insisted the Bass Strait price should not be allowed to float with world values as the government had suggested.

As is now well documented, the Bass Strait producers were wrong. The world price rose sharply during the mid-1970s, and Esso/BHP were caught in a fixed price agreement of their own making. Despite appeals to successive governments in Canberra in the early 1970s, there was no change in pricing policy until late 1975, when the Whitlam adminstration added twenty-three cents a barrel to the previous Bass Strait price. During that five year period (1970-1975) exploration offshore Gippsland began to languish as drilling costs rose and incentives fell away.

Esso and BHP were bitterly disappointed with the Whitlam announcement as it did nothing to redress the imbalance with world values. Whitlam did, however, introduce the policy of world parity pricing for any finds made after 1975. It is impossible to know what the outcome of this 1975 policy would have been, because a few months afterward Whitlam and his Labor colleagues were bundled out of office and Malcom Fraser became Prime Minister.

The Fraser oil policy was not clarified until the Federal Budget of 1977. It imposed a complicated system of levies on Bass Strait which put a gradually increasing part of the production at world levels (about $A12.60 a barrel at that time) and left the bulk at $A2.33 a barrel. But perhaps the most significant factor for Esso/BHP's exploration incentives was that Fraser retained Whitlam's policy of parity pricing for any new finds, and promised that they would be levy free.

The first test came with Fortescue in 1979. When this field was granted 'new oil' status (co-inciding with the Iranian revolution and renewed world oil price rises) Bass Strait activity came to life for the first time in a decade. Australian parity prices rose to over $38 a barrel by the end of 1982. The policy remained relatively unchanged until 1983, when the Fraser Administration was toppled by Labor under Bob Hawke.

Hawke came to power promising changes, including replacement of the levy/new oil system with a resource rent tax which would be a tax on oil profits and not on production. But after many discussions Esso/BHP managed to convince the Labor Government that it would be virtually impossible to impose RRT on existing Bass Strait production, and impractical to place such a regime on future production. Hawke conceded those points and made only minor changes to the 'old' oil (oil discovered pre-1976) levies. However he did withdraw the levy free status of 'new' oil, thus placing a royalty on Fortescue for the first time.

Nevertheless the parity base price was retained at around $A35 a barrel and was still attractive for any small fields of less than 3000 barrels a day production rate.

The most recent change (October 1984) was the creation of yet another tier in the levy system — this time an intermediate step between the 'old' oil and 'new' oil — which gave Esso/BHP the economic incentive to plan development of previously marginal fields. Hence the announcement of the Strait's fourth development program.

This has now begun with a $400 million platform construction project for tapping the 35 millions barrels of oil plus gas reserves at Bream. Unfortunately the 1986 price crash has caused postponement of the rest of Esso/BHP's plans. They can, however, be reactivated rapidly when the oil price improves.

Looking back Bass Strait has been the country's mainstay in oil for the past sixteen years, and yet it is interesting to note that the greatest production rate came during 1985, when an average of 500 barrels a day flowed from the Esso/BHP fields.

This came about after engineers worked hard to 'debottle-neck' the production system in response to price incentives and export permits. The figure of 550 000 barrels a day was reached and surpassed more than once in the span of that twelve months.

The first-ever Australian crude oil exports from the region began in 1984 and continued throughout 1985 into 1986. The highest revenue totals were recorded during 1985 with an average of 160 000 barrels a day of Bass Strait production marketed overseas mainly in south-east Asia and the USA west coast.

But the major question now is how long can this continue into the future? It is probable that the 1960s-style excitement of big discoveries has gone forever, but the potential for albeit smaller finds and the engineering challenge of difficult new development is still present. There is no doubt that Bass Strait gas reserves will exceed Victoria's needs for the rest of this century. An equally arguable point of view is that the classic Tertiary age reservoirs will also supply much of Australia's oil production for the same period.

**References:**

Australian Institute of Petroleum: *Petroleum Gazette*,
    September 1978, September 1984

DA Brown, KSW Campbell and KAW Crook: *Geological
    Evolution of Australia and New Zealand*, Pergamon
    Press, Sydney 1968

Susan Bambrick: *Australian Minerals and Energy Policy*, Australian National University Press, Canberra 1979

Donald W Barnett: *Minerals and Energy in Australia*, Cassell, Australia 1979

CEB Conybeare: *Oil Search in Australia*, Australian University Press, Canberra 1980

Robert Murray: *Fuels Rush In—Oil and Gas in Australia*, Sun Books, Melbourne 1972

JRV Prescott (ed): *Australia's Continental Shelf*, Nelson, Melbourne 1979

HG Raggatt: *Mountains of Ore—Mining and Minerals in Australia*, Landsdowne Press, Sydney 1968

Lewis G Weeks: *A Lifelong Love Affair. The Memoirs of Lewis G Weeks, Geologist*, Wisconsin University Foundation, Madison, USA 1978

Rick Wilkinson: *A Thirst For Burning—The Story of Australia's Oil Industry*, David Ell Press, Sydney 1983

# 11
# BASS STRAIT: LIGHTHOUSES AND WRECKS

## Stephen Murray-Smith

*About midnight every second Tuesday night all the year round, Frank Goold tumbles out of bed in his Port Albert home, sleepily walks out to his car and drives down to the windswept little jetty. There, snugly tethered, is his 48-foot fishing boat with a name that is heard more often on the Bass Strait airwaves than that of any 20 000-ton liner: the* Marjorie Phyllis. *In the wheelhouse Frank switches on his wireless transmitter and calls 3GR, 3GS and 3GT. His voice goes crackling out down south, over the heaving white-capped swells of Bass Strait and over the little granite islands. Cliffy Island and Wilson's Promontory and Deal Island are waiting for him. Four or five days ago they wirelessed in their orders, and a couple of dozen human beings are waiting for their Scotch fillets, their bedroom slippers, their cigarettes, their books and magazines, their letters and their beer. Now Frank says to them: Can I get out? Can I work Cliffy, or is the easterly swell still coming in on the reef there? What's it like down south of the Prom., on that long six-hour leg he has to run down to Deal Island in the middle of the Strait?*

*Chances are all will be well. Anxious though they are to see him, the lighthouse people won't let this colour their weather reports. But they know it takes a lot to hold Frank back, unless he's got passengers, women and children specially, whom he'll protect from a gruelling trip if he can. So with Frank it's back to bed for a couple of hours, then back to the wharf again. A trip to the freezer, to load the meat and other perishables. Then, three o'clock or not much later, cast off and down the channel. Forty minutes or so, and it's over the bar—a tricky business this, even in daylight. By sunup he is well on his way to his first stop, Cliffy Island, three hours off the Port.*

*Cliffy's just an eleven-acre lump of rock, sticking up 160 feet out of the sea. On top of it is the light itself, and three houses for the three families of this watch-keeping station—that means that someone's on duty at all hours, day and night. Cliffy is not the most isolated, but must be one of the most confined stations around the Australian coast. To land people or stores here*

*Cliffy Island light-station in Bass Strait—not far from Port Albert—is now demanned. It was known to light-keepers as 'the Alcatraz of Bass Strait'—just a rock.*
*Original housing compounds, with rock walls to keep out the winds, are visible.*
*Photo: Marc Fallander*

*requires seamanship and nerve on everyone's part. From a platform near the top of the cliff, a dinghy with two men on board is lowered from the derrick into the water, and motors out cautiously through the swell and the reefs to where the Marjorie Phyllis is lying as close in as she can get. Stores and mail are manhandled over the side, there are hurried greetings and perhaps a shouted item of news, and then the dinghy goes back up the cliff while the ketch steams off on its next leg down to the Prom.*

*Although the Promontory lighthouse is on the mainland, there's no road through to it, and from the point of view of the lighthouse people it's like living on an island. There's no jetty and the landing is by seizing your chance as the swells rise up on the smooth rocks and glissade back again. After this Frank is out of the lee of the Prom., driving out into the Strait itself. It's about morning-tea time, and if it's not too rough he'll make himself a cuppa. Out of his wheelhouse window, during the next few hours, he'll be able to observe a grand frieze of craggy islands slowly passing at eight knots: Rodondo, the Moncoeurs, the Devil's Tower and Sir Roger Curtis's Island. He'll be steering a compass course for the Kent Group, but more likely than not, if it's a reasonably clear day, there'll be a little cloud sitting above the Group, and all the other islands too, like a genie above a lamp—and these are handy things to steer for when you can't see the land itself. He'll be among the mutton-birds now, if it's the mutton-bird months, and among the great albatrosses and the little petrels too—the petrels that daintily dance along the wave tops and which got their name that way, after Saint Peter. The whales, and all that move in the Waters, are infinitely closer to you in a small boat, with a freeboard of two or three feet, than from the promenade deck of an ocean liner.*

*Three or four hours off his destination, Frank will pick up the Group. Slowly the peaks and the cliffs, the bays and the different coloured greens will become clear. Soon he'll be nosing in around the south of Dover Island, around those frightening cliffs the fishermen call 'Mother-in-law's Leap', and into the sheltered waters of Murray Pass. Five minutes later and he'll be anchored in Brown's Bay and taking his hatch covers off. Bob McNeill, the headkeeper, will already have the station dinghy in the water, Tracey and Geoffrey and the other lighthouse kids will be jumping up and down at the end of the jetty. It's the end of a long trip for Frank Goold, but it's not at all sure that he'll accept the invitation to come ashore for a feed and a yarn. If there's any sign of the weather worsening—*

*Wilson's Promontory lightstation. Matthew Flinders called Wilson's Promontory 'cornerstone to a continent'.*

*Alterations to the lighthouse at Wilson's Promontory. This light commands the main shipping channel, known as 'Bourke Street', to the keepers, and is a crucial turning point for traffic through the straits in and out of Melbourne.*

*and Bass Strait weather seems to be worsening about ninety-eight per cent of the time, if that is mathematically possible—he'll be off again on his nor'nor'west course home within minutes, and back at Port Albert eight hours later after nearly twenty-four hours on his feet.*

Well, that was the way it was when this book was first published in 1969. Cliffy Island was 'automated' in 1971. Deal Island was supplied by the 'ketch' until 1978. (It wasn't a real ketch, of course, but that was the old Bass Strait term, and it survived until the end, and 'ketch day' was the fortnightly red-letter day when supplies and, sometimes, visitors arrived at the isolated bass Strait lightstations.) Many were sad when the *Marjorie Phyllis* was sold and disappeared from Bass Strait waters. Frank Goold, now (1985) seventy-three years old, lives in active retirement with Flo near Port Albert, and is a demon (lawn) bowler. His famous cry as the *Marjorie Phyllis* bucketed over the waves, 'Anyone for a banana sandwich?' is no longer heard in the land.

Today we start such a piece as this in a different way. At a quarter to six every second Tuesday Peter Clemence throws back the blankets and gets out of his bed at Mount Eliza, on the Mornington Peninsula. He breakfasts, pulls on a helmet, mounts his Honda Twin 400 motor-bike and rides up the freeway to Moorabbin airport. At the Jayrow helicopter office he assesses the latest weather information, fills out a flight-plan, checks if there are any parcels for Wilson's Promontory or Deal Island, and takes off in his Bell Jet-Ranger at 8 am. He may have a technician with him to put in a day's work at the Prom light-station, and if so he flies straight there, over the green cow-cocky country of south Gippsland, the rolling Strzeleckis flattened out from the two or three thousand feet at which he is flying, and then the dramatic granite mountains and the long, deserted beaches of Wilson's Promontory.

If there are no passengers Peter flies directly to the little airfield at Port Welshpool. At Welshpool he gets out a hand-operated pump and refuels from drums of special fuel stored there. Waiting for him is the truck from the Moorabbin work-shops of the Department of Transport. The truck has detoured to Foster on the way down over the hills in order to pick up groceries and other items of a personal nature the lightkeepers at Deal Island and the Prom have ordered by phone a few days beforehand. It's a tiny cabin, and loading the helicopter with packages in all shapes and sizes is a special art. Then he's off, up and over the coastline, round Mount Singapore on the Prom and setting his course south-east for the Hogans and then for the cliffs of the Kent Group. Sixty-four nautical miles and fifty minutes later Peter is circling round the light, and is coasting gently down in an elegant curve to the helicopter pad in the compound, just by the 1847 'historic quarters', a house that has seen so much Bass Strait history.

The keepers, Stan and Shirley Grey and Max and Linda Lucas, are waiting there for him. Peter says he never fails to be delighted with the welcome he gets at these stations. Unloading, perhaps a quick cup of coffee, off again, through the Funnel between Erith and Dover Islands, over the sea streaked with white to the tiny helipad perched on the edge of the cliff at the Prom — sometimes the attempt to land must be abandoned — to another welcome, from Ted and Anne Peers and Steve and Pam Du Bois. Off again, back to Welshpool for another load for one or other of the stations, and so through a busy day until the sun is low and it's time to head for home, leaving behind Bass Strait granite for, as Peter says, 'the soft green undulations of the Strzelecki Range as it rolls away to the north-east in the afternoon light'. (There may not be poets out there, but there is poetry, and I remember Frank Goold once, coming into Erith Island after dark and after a hard trip, saying 'After the Hogans the sun set, and the wind went away with the sun'.) At dusk Peter finds himself back at Moorabbin, glad to shut down engines, a long day of sea, wind, islands and fine judgements behind him.

*

I am very fond of the Deal Island light. For a start, it's spectacularly beautiful, perched up there a thousand feet above sea level — though unfortunately it's often hidden by cloud, as it has been for the last 138 years, and smaller supplementary lights are being installed on islands around the Kent Group. Archives tell us that the Deal Island light was built with immense labor, using bullock-teams and convict workers, back in 1847; home for many years in the nineteenth century of families like the Baudinets and the Browns, large families, too, born and brought up in this remote environment where, if the three-monthly supply vessel from Hobart Town was a month or two overdue, the lusty boys of the family would think little of pulling in the whaleboat five or six miles offshore to hail a becalmed ship for news and perhaps some badly needed supplies. The old

*Original quarters of Superintendent, Deal Island lightstation, Kent Group. These historic quarters were built by convict labour in 1847 and were visited by many distinguished people in the 19th century on 'missionary visits'.*
*Photo: Andrew R. Lyell.*

colonial cottage on Deal Island, with its steep staircase and its little attic rooms, the cottage which housed the head keeper and his family for the best part of a century, is preserved by the lighthouse adminstration as what they call the 'historic quarters'.

I have mentioned three of the lights in Bass Strait, the three that form a triangle east and south of Wilson's Promontory. These are three of the oldest lights, but not the very oldest: this honour belongs first of all to Low Head, at the mouth of the Tamar, which dates from 1833 and which is Australia's third-oldest lighthouse; and then to the lighthouses on Goose Island, which lies to the west of Flinders Island, and on Swan Island, off Cape Portland on the north-east of Tasmania. Both were built in the mid-1840s, a couple of years before the Deal Island light, and like the Deal Island light, and the Otway light, and the Schanck light, and the Point Hicks light, and the Nelson light, and the Prom. light, and the great red granite Gabo light, off Mallacoota, and like the even taller Cape Wickham light of 158

120

*The great thousand-foot cliffs of Deal Island, with the 1847 lighthouse, a signal to mariners for 150 years.*
*Photo: Marc Fallander*

Head-keeper Stan Grey polishes the lenses of his Deal Island lighthouse,
at a thousand feet above sea level said to be one of the two most elevated
in the world.
Photo: Marc Fallander.

feet (50 m), the tallest in Australia, on the north end of King Island—like all these sturdy granite constructions, they look good for hundreds of years yet.

One interesting thing historically about the Bass Strait lighthouses is that building them and maintaining them first drew the Australian colonies together. Between New South Wales and Van Diemen's Land in the first place, and then after 1851 between New South Wales, Victoria and Tasmania, there was little love lost. But on certain essentials, and the lighthouses were the first of these, they just had to find ways of co-operating. You can easily trace how it happened. By 1840 the colonies had passed out of their essentially convict period. They were becoming new Britannias in another world. International trade and commerce was increasing. Down in Hobart Town, Sir John Franklin took up his pen—the John Franklin who, as a midshipman forty years before, climbed Arthur's Seat with Matthew Flinders, the Franklin who was to perish in the northern ice within a few years of writing his letter to Sir George Gipps in Sydney, the Franklin whose name is borne by a big fishing vessel from Hobart I have seen drop anchor in West Cove in the Kent Group, and by the new CSIRO oceanographic vessel, based at Hobart. 'The prevalence of strong winds,' Franklin wrote, 'the uncertainty of either the set or force of these currents, the number of small rocks, islets, and shoals, which . . . have been but imperfectly surveyed, combine to render Bass' Strait under any circumstances an anxious passage for a seaman to enter; and I will venture to say that the master of any merchant ship trading from the northern hemisphere to Sydney, or in the opposite direction, looks upon this portion of his voyage with the greatest apprehension.'

The mariners and the merchants of Australia were already concerned about the wrecks and the near-misses in Bass Strait. But an event of August, 1845, was to shock the whole Australian community. The *Cataraqui* was a ship of 800 tons, out of Liverpool for Melbourne, Captain Finlay, master. A crew of forty-six and 369 emigrants, largely marriageable girls for lonely colonists, but including seventy-three children. The seas were monstrous, no observations had been possible for some days. Despite arguments, the captain was obstinate. The *Cataraqui* ran before the gale and at 4.30 am, without even a warning cry of 'breakers ahead', crashed and crunched onto the west coast of King Island, where her remains lie to this day with those of sixty other vessels. A couple of days later David Howie, who bore the title of Constable of the Straits, found 401 bodies on the beach, and nine survivors. The dead were buried in four mass graves. To this day it remains Australia's greatest civil disaster. This appalling event, and others nearly as bad, made our forefathers come together to save life and promote profits.

There have always been wrecks in Bass Strait. When the first white men came they found mysterious wrecks, wrecks which no one has ever been able to account for. There have been many famous wrecks, like the *Admella* and the *Loch Ard*, and some not so famous, like the *Australia* on Corsair Rock, after which my grandfather found himself stepping out of the train at Spencer Street in his pyjamas. Every year there are wrecks, when some fishing boat skipper gets overconfident of his skill, or relies too much on his automatic pilot. And the treachery of the Bass Strait waters is proverbial.

The wrecks brought the lighthouses then, but the lighthouses also brought a few wrecks, especially when, at the end of their long Great Circle run down through the Southern Ocean, the skippers mistook the Wickham Light on King Island for the Cape Otway light. If you glance at your map you'll quickly see what *that* involved. But the lighthouse service is probably Australia's best public investment. It is an unassuming service, the *real* silent service.

Since the first edition of this book the Australian lighthouses have achieved a new prominence. Towards the end of the 1970s the Commonwealth Department of Transport developed plans to turn most, if not all, of the forty-odd manned Australian lightstations into automatic stations, bereft of their keepers. The motivation was quite simple. Under the Lighthouses Act the sole responsibility of the Department of Transport was to keep a light burning, and modern technological developments, which included sophisticated alarm systems if something went wrong, made it possible to do this without the daily rigmarole of keepers visiting the towers, drawing the curtains, switching on the power — with, of course, the reverse procedure at dawn.

The government plans aroused a storm of opposition and led to the formation of a community body to defend the lightstations, the Australian Lighthouse Association. This association pointed out that, whatever the Lighthouses Act said, in practice the duties of lightkeepers went far further than keeping a light burning. Over the past century and a half many extra responsibilities had accrued: weather reporting, search and rescue liaison, coast-watching, the protection of both built and natural environments which were increasingly precious, just because they *were* lightstations and had been protected from

*Grim warning of the danger of Bass Strait: the fishing vessel WILLIAM FLAIR on the rocks of Prime Seal Island in 1976.*
*Herald and Weekly Times*

development for so long. No automatic apparatus, it was pointed out, could tell a Port Albert fisherman, ringing into Deal Island before he left on a voyage, the state of the sea in the centre of Bass Strait. And manned lightstations certainly did, from time to time, save lives. How was this to be put into a financial profit-and-loss account?

The Department of Transport admitted the opposition case, but stated that the 'accrued' responsibilities of lightkeepers were properly the financial responsibility of others. Out of this came a parliamentary enquiry in 1983, and from that has emerged a recommendation that the considerable majority of Australian manned lightstations, perhaps thirty-three of them, will remain that way. Consultations with community bodies and State governments and between Commonwealth departments are taking place to spread the responsibilities of the continuance of a human presence at these outposts. It all appears to have been something of a triumph for common sense, though one may well ask how we got so far from common sense in the first place.

Bass Strait will continue to be a place which wise men and women will treat with respect. All the more so because, in recent years, it is increasingly a venue for recreational sailing and boating. The oil rigs have brought a continuous naval presence in the straits, engaged in the sometimes hair-raising task of preventing foolish ship-masters from ramming the rigs. But soon, within twenty years or so, the rigs will vanish. Bass Strait will remain, turbulent and, for many, tragic. The lights will shine from their white towers, and all who see them will receive the message that, after all the idiocies of our behaviour are taken into account, human beings still care for each other.


*Footnote*
Since January 1986 the helicopter services to Wilson's Promontory and Deal Island have been carried out by Ron and Brett Newman, of Professional Helicopter Services Pty. Ltd.